Predictive Analytics Using Statistics and Big Data: Concepts and Modeling

Edited by

Krishna Kumar Mohbey
Central University of Rajasthan
India

Arvind Pandey
Department of Statistics
Central University of Rajasthan
India

&

Dharmendra Singh Rajput
VIT Vellore
India

Predictive Analytics Using Statistics and Big Data: Concepts and Modeling

Editors: Krishna Kumar Mohbey, Arvind Pandey and Dharmendra Singh Rajput

ISBN (Online): 978-981-14-9049-1

ISBN (Print): 978-981-14-9051-4

ISBN (Paperback): 978-981-14-9050-7

© 2020, Bentham Books imprint.

Published by Bentham Science Publishers Pte. Ltd. Singapore. All Rights Reserved.

need for a court order if at any point you breach any terms of this License Agreement. In no event will any delay or failure by Bentham Science Publishers in enforcing your compliance with this License Agreement constitute a waiver of any of its rights.

3. You acknowledge that you have read this License Agreement, and agree to be bound by its terms and conditions. To the extent that any other terms and conditions presented on any website of Bentham Science Publishers conflict with, or are inconsistent with, the terms and conditions set out in this License Agreement, you acknowledge that the terms and conditions set out in this License Agreement shall prevail.

Bentham Science Publishers Pte. Ltd.
80 Robinson Road #02-00
Singapore 068898
Singapore
Email: subscriptions@benthamscience.net

CONTENTS

FOREWORD

Big data analytics started receiving increasing attention a few years ago. It all began by exploring how to deal with the increasing volume, variety, and velocity of the data. Today, developing effective and efficient approaches, algorithms, and frameworks are used as they are essential to deal with such data.

Earlier, data were limited, and experiments were also performed for limited scopes. With the advent of big data, internet technologies, massive information, and predictive analytics problems have exploded in complexity and behavior. There are increasing challenges in the area of data storage, management, and computations. There is a need to combine various researches related to big data technologies, statistics, and predictive analytics into a single volume.

The proposed eBook addresses a comprehensive range of advanced topics in big data technologies with statistical modeling towards predictive analytics. This book will be of significant benefit to the community as a useful guide of the latest research in this emerging field, *i.e.*, predictive analytics. This ebook will help the studies in this field finding relevant information in one place.

R.S. Thakur
Maulana Azad National Institute of Technology
Bhopal
India

PREFACE

Predictive analytics is the art and science of proposed predictive systems and models. With tuning over time, these models can predict an outcome with a far higher statistical probability than mere guesswork. Predictive analytics plays an essential role in the digital era. Most of the business strategies and planning depend on prediction and analytics using statistical approaches. With the increasing digitization day by day, analytical challenges are also increasing at the same rate—digital information, which is rapidly growing, generating vast amounts of data. Hence, the design of computing, storage infrastructures, and algorithms needed to handle these "big data" problems. Big Data is collecting and analyzing complex data in terms of volume, variety, and velocity. The most extensive selection of big data is from digital information, social media, IoT, sensor, *etc*.

Predictive analytics can be done with the help of various big data technologies and statistical approaches. Big data technologies include Hadoop, Hive, HBase, and Spark. There are numerous statistical approaches to perform predictive analytics, including Bayesian analysis, Sequential analysis, Statistical prediction, risk prediction, and decision analytics.

This book presents some latest and representative developments in predictive analytics using big data technologies. It focuses on some critical aspects of big data and machine learning and provides descriptions for these technologies.

The book consists of seven chapters. Chapter 1 discusses data analytics in multiple fields with machine learning algorithms. An application of bootstrap sampling is presented in chapter 2 with the case study of quantifying player's monopoly in a cricket team. Successive sampling for mean estimation is discussed in chapter 3. Chapter 4 discussed log type estimators of population mean under ranked set sampling. Bivariate survival data analysis is represented in chapter 5. An approach for weblog data analysis using machine learning techniques is discussed in chapter 6. Chapter 7 discussed an epidemic analysis of COVID-19 using exploratory data analysis approaches.

Many eminent colleagues made a significant impact on the development of this eBook. First, we would like to thank all the authors for their exceptional contributions to the eBook and their patience for the long process of editing this eBook. We would also like to thank the reviewers for their insightful and valuable feedback and comments that improved the book's overall quality.

Krishna Kumar Mohbey
Central University of Rajasthan
India

Arvind Pandey
Central University of Rajasthan
India

&

Dharmendra Singh Rajput
VIT Vellore
India

List of Contributors

Anoop Kumar	Department of Mathematics and Statistics, Dr. Shakuntala Misra National Rehabilitation University, Lucknow, India
Arvind Pandey	Department of Statistics, Central University of Rajasthan, India
Bireshwar Bhattacharjee	Department of Statistics, Assam University, Silchar, Assam, India
Brijesh Bakariya	Department of Computer Science and Engineering, I. K Gujral Punjab Technical University, Punjab, India
Chemmalar Selvi G.	School of Information Technology and Engineering, VIT University, Vellore, India
Dharmendra Singh Rajput	VIT Vellore, India
Dibyojyoti Bhattacharjee	Department of Statistics, Assam University, Silchar, Assam, India
Krishna Kumar Mohbey	Central University of Rajasthan, India
Lalpawimawha	Department of Statistics, Pachhunga University College, Mizoram, India
Lakshmi Priya G.G.	School of Information Technology and Engineering, VIT University, Vellore, India
Nishi Rastogi	Department of Statistics, National P. G. College, Lucknow, India
R. Suguna	Sri Sarada College for Women (Autonomous), Salem-16, Tamilnadu, India
R. Uma Rani	Sri Sarada College for Women (Autonomous), Salem-16, Tamilnadu, India
Shailja Pandey	Department of Mathematics and Statistics, Dr. Shakuntala Misra National Rehabilitation University, Lucknow, India
Shashi Bhushan	Department of Mathematics and Statistics, Dr. Shakuntala Misra National Rehabilitation University, Lucknow, India
Shikhar Tyagi	Department of Statistics, Central University of Rajasthan, India

CHAPTER 1

Data Analytics on Various Domains with Categorized Machine Learning Algorithms

R. Suguna*, R. Uma Rani

Sri Sarada College for Women (Autonomous), Salem-16, Tamilnadu, India

Abstract: Data Analytics is an emerging area for analyzing various kinds of data. Predictive analytics is one of the essential techniques under data analytics, which is used to predict the data gainfully with machine learning algorithms. There are various types of machine learning algorithms available coming under the umbrella of supervised and unsupervised methods, which give suitable and better performance on data along with various analytics methods. Regression is a useful and familiar statistical method to analyze the data fruitfully. Analysis of medical data is most helpful to both patients as well as the experts to identify and rectify the problems to overcome future problems. Autism is a brain nerve disorder that is increasing in the children by birth due to some most chemical food items and some side effects of other treatments and various causes. Logistic Regression is one of the supervised machine learning algorithms which can operate the dataset of binary data that is 0 and 1.

Agriculture is one of the primary data which should be considered and analyzed for saving the future generation. Rainfall is a more elementary requirement for the global level and also countries which are having backbone as agriculture. Due to the topography, geography, political, and other socio-economic factors, agriculture is affected. Thus, the demand for food and food products is intensifying. Especially crop production is depending upon the rainfall, so, prediction of rainfall and crop production is essential. Analysis of social crime relevant data is indispensable because analytics can produce better results, which leads to reducing the crime level. Unexpectedly child abuse is increasing day by day in India. Linear regression is the supervised machine learning algorithm to predict quantitative data efficiently.

This chapter is roofed with various datasets such as autism from medical, rainfall, and crop production from agriculture and child abuse data from the social domain. Predictive analytics is one of the analytical models which predict the data for the future era. Supervised machine learning algorithms such as linear and logistic regression will be used to perform the prediction.

***Corresponding author R. Suguna:** Sri Sarada College for Women (Autonomous), Salem-16, Tamilnadu, India; Tel: +91-4274550291; E-mail: sugunarmca@gmail.com

Keywords: Data analytics, Exponential distribution, Inomial distribution, Linear regression, Logistic regression, Machine learning, Normal distribution, Prediction analytics.

INTRODUCTION

Data Analytics

Analysis of data is a need and essential task to get the solutions for the problems. In our society, there is plenty of progress going, and data are increasing day by day. Analytics helps to analyze the data. There are various real kinds of domains, such as medical business, agriculture, and crimes, which are the most considerable areas. Analytics is advanced mining of data with the standard statistical and mathematical methods. There are various kinds of analytical methods available, they are,

- Descriptive Analytics
- Diagnostic Analytics
- Predictive Analytics
- Prescriptive Analytics

Descriptive and diagnostic analytics concentrated on past events, which analyze the past data. Predictive and prescriptive analytics are used to give future solutions for past data [1].

Machine Learning

Fig. (1) shows the classification of machine learning algorithms. Machine learning is apart from the root of artificial intelligence. A machine can read and analyze the data. There are some wings of machine learning that are supervised, unsupervised, and reinforcement learning.

Supervised Machine Learning

Classification

Classification is the process of classifying the data with the pre-specified rules. The classifier's labels are predetermined. Prediction can be made using the classification and regression techniques. There are various famous machine learning algorithms available for classification. They are listed below.

Random Forest

- It is an ensemble method, and it reduces the over-fitting of the result. This algorithm is mainly used for prediction. From the training data, voting is used to get the prediction.

Decision Trees

- Decision trees are another effective method for classification. It is a tree-like model consisting of root and leaf nodes. It can handle high dimensional data and provides accuracy at a reasonable level.

Nearest Neighbour

- It is a simple classification algorithm that classifies the nearest samples based on the likeness of the data points. Choosing the k value is most important in this algorithm. Based on the cross-validation or else square root of the samples, k value will be better on the training data.

Boosted Trees

- Boosting is a process of making a compilation of adapting weak learners to powerful learners. It is an ensemble method, and the weak learner is a classifier that is scarcely correlated with classification, and a powerful learner is a classifier with a strong correlation.

Support Vector Machine

- SVM is a robust classification algorithm that classifies the data on the N-dimensional space, and N is the number of features. SVM will find a hyperplane to classify the data. The hyperplane is a boundary for the classification of data.

Neural Networks

- The neural network is working based on the biological neuron. The layers of neural networks are the input layer, an output layer, and the hidden layer. The input layer takes the samples of input data, and processing of data is done by the hidden layer, then the calculated output is produced by the output layer.

Naive Bayes

- Naive Bayes is targeting the text classification, and it is working based on the conditional probability method. This algorithm produces classification trees based on the probability of the data. This tree is also known as a Bayesian network [2].

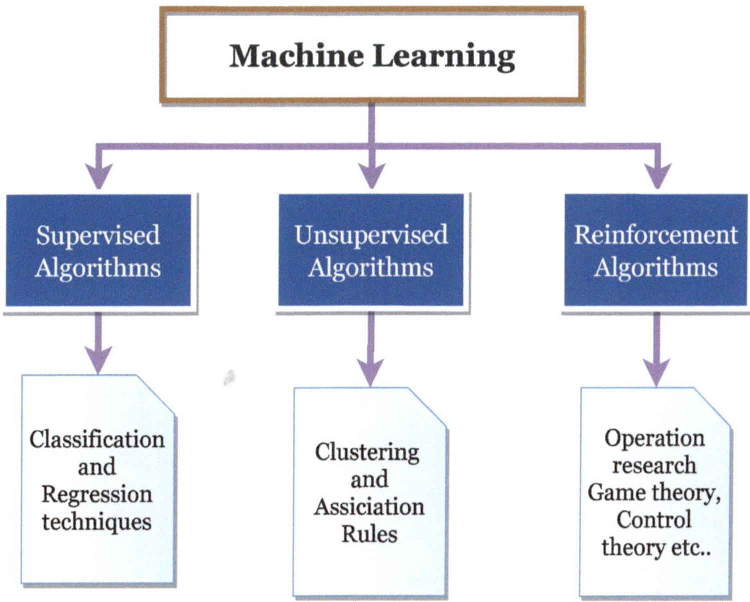

Fig. (1). Types of machine learning algorithms.

Regression

Regression is the prediction technique, which is the association among the dependent and independent variables. Regression plays an essential role in modeling and analyzing the data. Fig. (**2**) shows various types of regression techniques. There are various types of regression techniques available. They are shown below.

- Linear regression is a linear relationship between the dependent and independent variables. It is working based on the conditional probability distribution.
- Logistic regression is a method to deal with the binary data, and it shows the relationship among the significant factor and predictors.

- Polynomial regression deals with the fitting of the least-squares method. It models the dependent variable based on the independent variable.
- Regression models fit with analytical representation by the stepwise regression method. It consists of some approaches, such as backward elimination, forward selection, and bidirectional elimination.
- Ridge regression is used to reduce the standard errors of the model. It deals with complex regression data.
- Lasso regression helps to decrease the inconsistency and to increase the exactness of the linear model. It is mostly used in feature selection and minimizes the coefficients to zero.
- Elastic-Net regression is a combination of ridge and lasso methods. It is helpful on multiple correlated features [3].

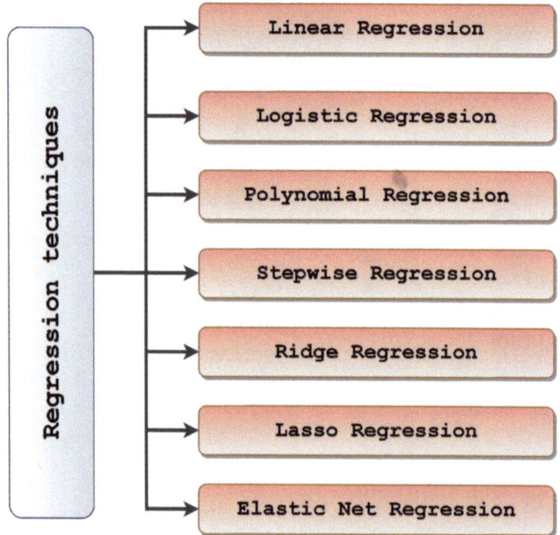

Fig. (2). Types of regression techniques.

Unsupervised Machine Learning

Clustering and Association Rules

- Clustering is a technique that handles the unlabeled data. Analogous data are grouping based on the similarities between the items. There are various algorithms available under this technique, *i.e*, K-Means, BIRCH (Balanced Iterative Reducing and Clustering using Hierarchies), DBSCAN (Density-Based Spatial Clustering), OPTICS (Ordering Points To Identify the Clustering Structure), Gaussian Mixture Model.

- Association rule describes the relationship between the exciting patterns. The terms such as support and confidence and lift are used in the association rule to get better patterns. The algorithms for the association rule mining technique are Apriori, AIS, SETM [4].

Reinforcement Learning

- Reinforcement learning is also called an approximate dynamic approach. It is applied in the robot control, elevator scheduling, *etc.*; Markov decision process is a suitable model in this learning method.

BACKGROUND OF DATA ANALYTICS

Analytics yields feasible solutions for the various domains of data such as Business, medical, agriculture, education, sports, and social problems. Cricket predictions were performed with the supervised machine learning algorithms such as decision trees, logistic regressions, naive Bayes, binary classifier model, and SVM. The dataset contains the cricket game details for the 19th and 20th centuries, then 3500 matches for training data and 1500 matches for testing data. Based on the comparison, the binary classifier model provided better performance among the other algorithms [5].

Nowadays, education sectors are searching for new technologies to improve the overall education field such as student performance, mentor enhancement, updation of technology for students, *etc.* Learning analytics has been discussed [6] to improve education. In the healthcare division, predictions of death due to the various causes were determined based on the dimensionality reduction, and classification algorithm has been applied to the data. The data belongs to the Queensland government. Most beautiful features selected with dimensionality reduction technique and SMORegression, MLPRegression, and Linear regression were used for prediction. Linear regression has given good results among the other algorithms [7].

Regression is an essential technique that gives an enhanced performance on the analysis of various types of data.[8]Analysed eye disease data with non-linear regression resolves the dark adaptation indicator in people. Parkinson's disease is one of the intimidation diseases around the world [9]. A Meta-analytic process has been applied to PD data for the duration of 1985 to 2010. Data belongs to

Americans, and 41 females and 134 males were classified as disease affected based on the algorithms SVM and adaptive boosting technique.

Agriculture is a particular domain which helps to increase economics and food production. There are ample problems available for the concentration, such as rainfall, soil, crop harvesting, crop production increasing, crop selection, pricing marketing distribution, import-export, and weather forecasting. Data analytics and machine learning techniques were used to identify and solve many of the problems of agriculture. The relationship between climate variability and crop yield was analyzed with regression technique [10]. Many of the authors grasped a variety of problems in agriculture, together with the statistical models and machine learning algorithms. The possible machine learning applications such as decision trees, ANN, KNN, neural networks, time series analysis, Markov chain model, and K-Means were discussed [11]. Climate change and variability in temperature of Ghana were analyzed with continuous distribution [12]. SVM, KNN, and least square method were applied for the crop estimation of sugarcane [13], seed classification for efficient crop production [14], android application for best profitable crops for appropriate weather conditions [15].

Autism spectrum disorder (ASD) is one of the exigent problems in society. Many of the children are affecting gradually, and parents are getting bothered about children's futures. ASD is not a chronic disease, but it is a damage to the brain nerve, and it affects children mostly. There are various types of ASD diseases, and text analysis has been presented [16] and various classification algorithms discussed, which are applicable to ASD. The classifiers such as logistic regression and support vector machine were applied and AUROC (Area under Receiver Operating Characteristic Curve) used to fit the model under cross-validation [17]. Another author [18] applied a Binary firefly algorithm for feature selection to get the optimum features on ASD data and applied various machine learning algorithms such as J48, KNN, SVM, naive Bayes. According to this work, SVM provided comfortable results among the other algorithms. The center for disease control prevention (CDC) America has organized a service demanding process to measure the pervasiveness of ASD. The work carried out [19] with eight supervised machine learning algorithms and random train test splits of data, and Naive Bayes-SVM has given high mean accuracy than other algorithms for the classification of ASD.

Child Abusing is a major social relevant problem. It is gradually happening even in rural as well as in the urban areas. Thus, the prohibition of physical abusing on children should be considered seriously in all the places. An analysis is needed and even takes over by the authors, along with the computationally intelligent methods. Analytics will be helpful in analyzing the data and avoiding it well-timed. Child abuse or maltreatment of children can be in the form of physical abusing, emotional abusing, sexual abusing, neglecting, *etc.* Risky behaviors among the youths have analyzed in Thailand with the measures of CTS (conflict tactics scales), diagnostic interview schedule (DIS), alcohol use disorder identification test (AUDIT), and linear regression were used to estimate the risk behavior outcomes [20]. An analytical approach of Predictive risk modeling (PRM) has developed by the team of economists of Newzealand to the center for applied research in economics. Stepwise regression was used in the data and AUROC applied to the data for the best fit of data [21].

Various Domains

In this chapter, an illustration is made with three types of data, autism from medical data, child abuse from social data, and rainfall data from agriculture. Autism consists of binary data, rainfall, and child abuse data consists of categorical data. Various types of regression methods will be applied to each data.

Medical Domain-Autism Data

Autism is a neurodevelopment disorder, and it commonly affects children up to 3 years. It is benchmark data shared by Fadithaptah, and 292 instances are available in this dataset. Physicians conducted various types of tests, such as learning, speaking skills, listening are conducted called as Child AQ-10 Test. The results are stored as binary values in the dataset.

Agriculture Domain-Rainfall Data

Rainfall is the most crucial factor in agriculture. The annual year outcome of crop production can be determined. Rainfall data has been collected from the Indian government website for the year 1901 to 2015. The rainfall level is stored in this dataset as millimeter measurement: month-wise, seasonal wise, and annual rainfall maintained in the dataset.

Social Domain-Child Abuse Data

Child abusing is an immense problem in society. Many children inflated due to this problem. Child abuse data is a benchmark data which consists of records of various states of India for the year 2001 to 2012. Different types of child abuse such as infanticide, foeticide, abetment of suicide, the murder of children, rape of children, kidnapping and abduction of children, buying of girls for prostitution, selling of girls for prostitution, the prohibition of child marriage act are recorded in this dataset.

ILLUSTRATION OF REGRESSION WITH VARIOUS DOMAINS

The overall flow of the chapter is described in Fig. (**3**). Regression is one of the statistical-based machine learning techniques which will provide fruitful results for data. There are different kinds of regression techniques that exist. The flow of the illustration with unlike data in this chapter will be in the following mode.

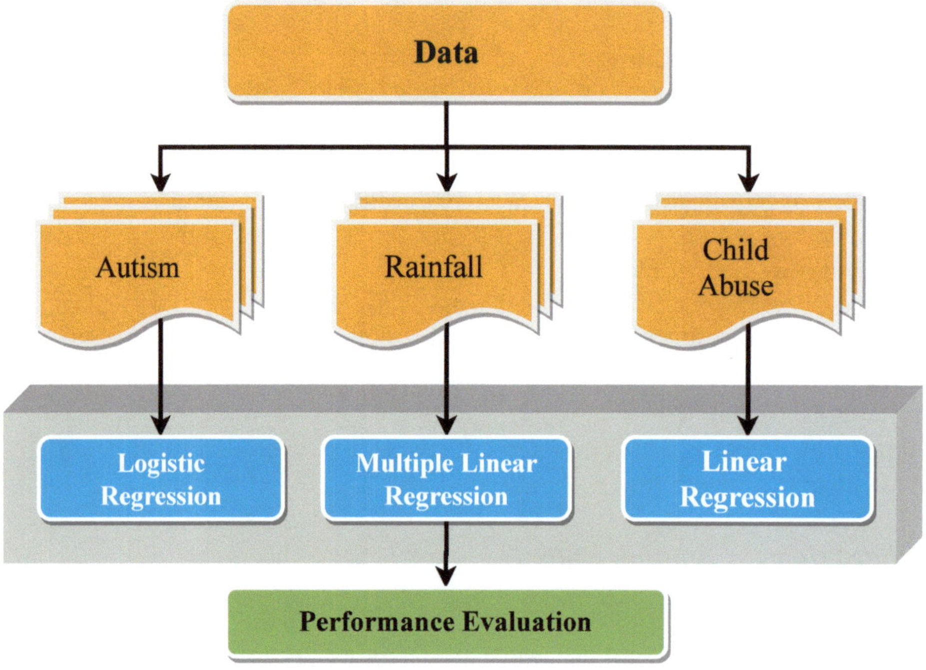

Fig. (3). Flow of the process with different kinds of data.

Logistic Regression

Logistic regression is another type which can give the prediction for binary outcome data. The method of logistic function makes the logistic regression. The logistic function is also called a sigmoid function, and it is an S-shaped curve that can take values and map this between 0 and 1. The logistic model description is,

$$F(y) = \frac{e^y}{1+e^y} \; for - \infty < y < \infty \tag{1}$$

Note that y->∞, f(y) ->1and as y->-∞, f(y) ->0.

Linear Regression

Linear regression is a simple and quiet better algorithm to make the prediction, which is used to model the relationship between the several independent variables and an outcome variable. It will provide the possible outcome value for the model. The linear model description can be expressed in the form of the following equation.

$$Y=\beta_0+\beta_1 x_1+\beta_2 x_2 ... +\beta_{p-1} x_{p-1}+\varepsilon \tag{2}$$

Where,

Y is a resultant variable

X_j is the input variables, for j=1, 2, 3...p-1.

B_0 is the value of y when each xi equals zero

B_j is the change in y based on a unit change in x_j, for j=1, 2...p-1.

ε is a random error term that represents the dissimilarity in the linear model and a particular experimental value for y [1].

Multiple Linear Regression

This technique is like linear regression, but it can work on more than two independent variables. It can apply to continuous and categorical variables. The model description is,

$$Y=\beta_0+\beta_1 x_{i1}+\beta_2 x_{i2}+...\beta_p x_{ip}+\varepsilon \tag{3}$$

For i=1, 2, 3...n. This model can be articulated as data=fit+residual, where fit symbolizes the term $\beta_0+\beta_1 x_1+\beta_2 x_2...\beta_p x_p$, and residual corresponds to divergence ε of experimental value y from their means μ_y. It is normally dispersed with mean 0 and variance σ [1].

RESULTS AND DISCUSSION

Logistic Regression for Autism Data

Table **1** shows a summary of the independent variables of autism data. The attributes, including age, test scores result for the children, and gender are taken for the account of the logistic regression function.

Table 1. Deviance Residuals.

Min	1Q	Median	3Q	Max
-1.5501	-1.1583	-0.7476	1.1712	1.7070

Based on Table **2**, Coefficients of the logistic regression function show that the hypothesis value is >0.5 with the 95% confidence level. Thus, the hypothesis H0: $\beta_{result} \neq 0$. The p-value for this model is 0.3427817 based on the chi-square test. The below picture shows that the positive class of autism occurs in the dataset, mostly.

Table 2. Model Description.

```
Coefficients:
              Estimate Std. Error z value Pr(>|z|)
(Intercept) -2.987e+02  8.611e+04  -0.003    0.997
gender       1.692e-01  1.623e+04   0.000    1.000
result       4.590e+01  1.316e+04   0.003    0.997
A1score      5.283e-01  1.410e+04   0.000    1.000
A2score     -5.073e-02  1.368e+04   0.000    1.000

(Dispersion parameter for binomial family taken to be 1)

Null deviance: 4.0446e+02  on 291  degrees of freedom
Residual deviance: 1.7190e-08  on 287  degrees of freedom
AIC: 10

Number of Fisher Scoring iterations: 25
```

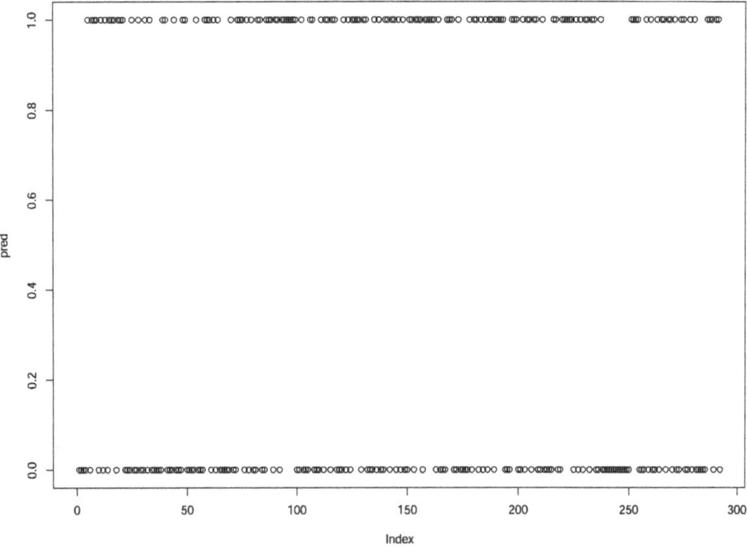

Fig. (4). Predicted Values of data.

ROC Curve

Fig. (**5**) shows that the true positive and false positive rate of logistic regression method. Here the AUC (Area under Curve) value is 0.74. It intimates the curve moving towards the top left corner. Negative values are predicted by this curve. Thus, the AUC value can be increased to get more performance.

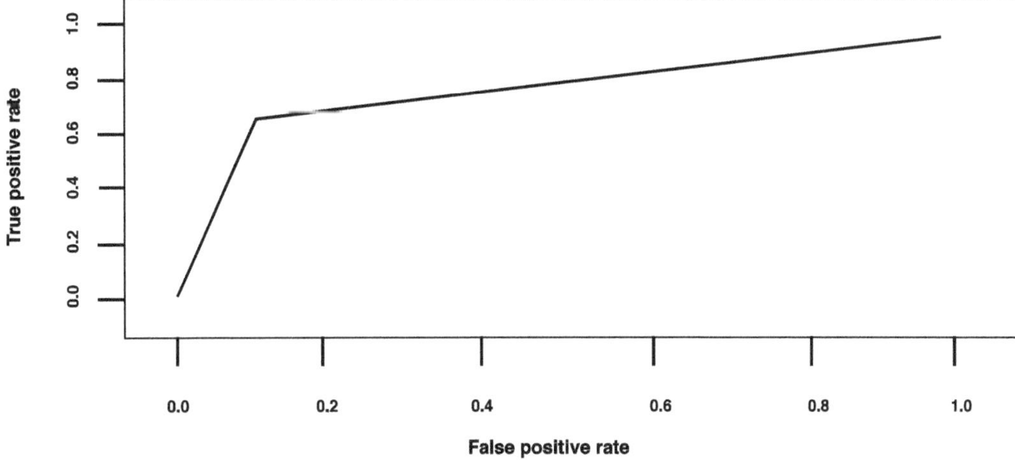

Fig. (5). Area under the curve for autism data.

Linear Regression for Child Abuse Data

Table **3** shows the residual summary of the linear model for child abuse data. The model's p-value and predictor's p-value are less than the significance level. R-squared and Adjusted R-Squared value should be higher. The model provided 0.9781, which is an excellent performance on data. The correlation among the data is nearby 1. Therefore, the data is positively correlated.

Table 3. Residuals of the linear regression model.

Min	1Q	Median	3Q	Max
-28369.8	-105.2	-104.2	-62.4	24112.4

```
Coefficients:
                Estimate Std. Error t value Pr(>|t|)
(Intercept)  105.18867  132.06988    0.796    0.426
crime$`2012`   6.28902    0.04243  148.220   <2e-16 ***
---
Signif. codes:  0 '***' 0.001 '**' 0.01 '*' 0.05 '.' 0.1 ' ' 1

Residual standard error: 2895 on 492 degrees of freedom
Multiple R-squared:  0.9781,Adjusted R-squared:  0.9781
F-statistic: 2.197e+04 on 1 and 492 DF,  p-value: < 2.2e-16
```

```
cor(crime$crimelevel,crime$`2012`)
[1] 0.9889871
conf<-predict(linearmod,newpt,level=.95,interval = "confidence"
)
>conf
             fit            lwrupr
1      142.9228  -116.486046      402.3316
2      105.1887  -154.301875      364.6792
3      105.1887  -154.301875      364.6792
4      117.7667  -141.696495      377.2299
5      136.6337  -122.788619      396.0561
```

According to the year 2012, crimes increase, then the overall crime level also increases. The predicted values for crime are calculated by using predict(). Lower level and upper-level crime range are displayed along with the fitted values.

Fig (**6**) explores the Q-Q plot and scatter plot for crime data. Q-Q plot shows the residual compares the observed data and against the quantiles of the distribution of data. The scatter plot displays the linear relationship between the crime level and the year 2012 data.

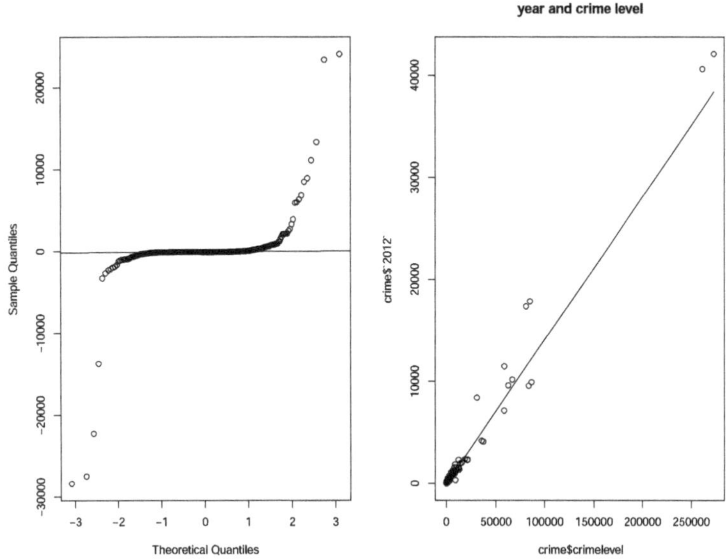

Fig. (6). Quantile comparison and the linear combination of data.

Table 4. Residuals of the MLR model.

Min	1Q	Median	3Q	Max
-0.218464	-0.062604	0.001647	0.075750	0.298330

```
Coefficients:
              Estimate Std. Error   t value Pr(>|t|)
(Intercept) -0.0548132  0.1149990    -0.477    0.635
JAN          0.9991073  0.0010190   980.506   <2e-16 ***
FEB          0.9991680  0.0008826  1132.029   <2e-16 ***
MAR          1.0009441  0.0008253  1212.884   <2e-16 ***
APR          0.9999059  0.0009800  1020.322   <2e-16 ***
MAY          1.0001094  0.0006625  1509.546   <2e-16 ***
JUN          1.0001078  0.0002923  3421.491   <2e-16 ***
JUL          1.0001807  0.0002610  3831.835   <2e-16 ***
AUG          1.0001001  0.0002990  3345.143   <2e-16 ***
SEP          0.9998956  0.0003213  3111.854   <2e-16 ***
OCT          0.9996650  0.0003969  2518.792   <2e-16 ***
NOV          1.0001170  0.0006158  1623.986   <2e-16 ***
DEC          1.0008260  0.0011643   859.563   <2e-16 ***
---
Signif. codes:  0 '***' 0.001 '**' 0.01 '*' 0.05 '.' 0.1 ' ' 1


Residual standard error: 0.1035 on 102 degrees of freedom
Multiple R-squared:      1,Adjusted R-squared:       1
F-statistic: 1.086e+07 on 12 and 102 DF,  p-value: < 2.2e-16
```

Multiple Linear Regression with Rainfall Data

Table **4** explores the result of multiple linear regression for the rainfall data. It provides R-Squared is 1; hence it is the well-fitted model for rainfall data. P-value is <0.05, and most of the variables are significant variables for the model.

(Fig. **7**) captures the standard error on the variable of rainfall data, normality, and dominant observations. The standard error reveals the accuracy level of the data, and residuals are the difference between the actual and observed values on the dataset. From the above plot, most of the values are positively distributed. Thus, the error is in the lower level.

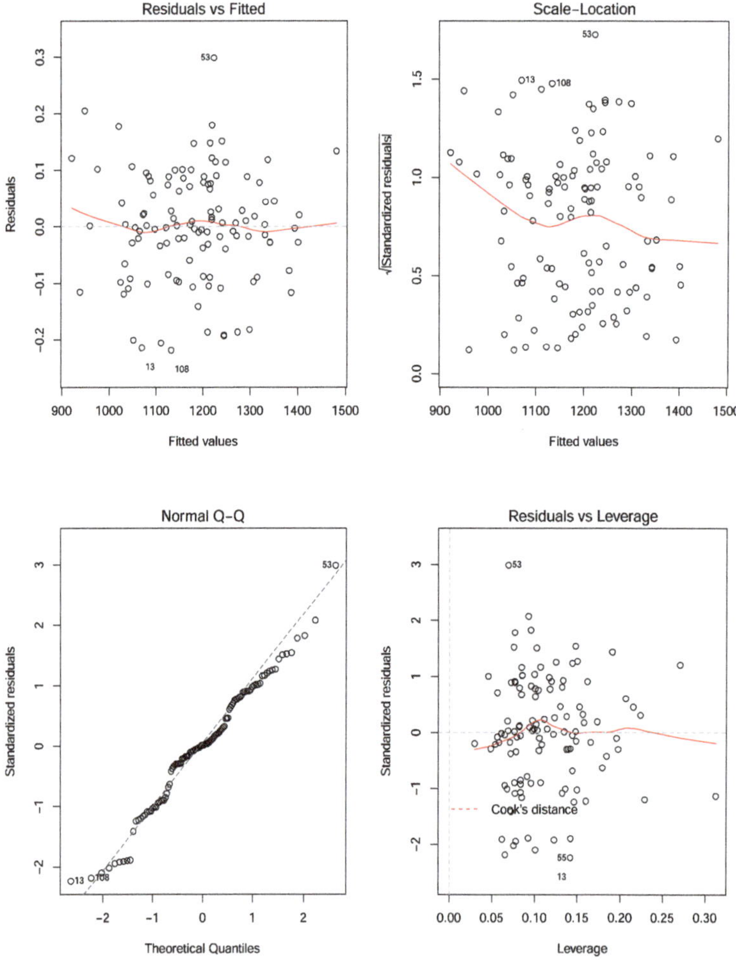

Fig. (7). Residual plots of MLR.

ANOVA (Analysis of Variance Table)

ANOVA is a statistical method that is useful to analyze the differences of mean between two or more groups in a dataset, as well as the equality of means. It is the shelter of the t-test. Based on Table **5**, the p-value is lower than the threshold value that is 0.05. Thus, the hypothesis between the variables is statistically significant.

Table 5. Model summary of ANOVA.

```
Model Summary
Response: ANNUAL

Df Sum Sq Mean Sq      F value     Pr(>F)
JAN           1  12090   12090   1128590.8 < 2.2e-16 ***
FEB           1  66024   66024   6163368.9 < 2.2e-16 ***
MAR           1     65      65      6090.6 < 2.2e-16 ***
APR           1  11202   11202   1045694.0 < 2.2e-16 ***
MAY           1 219104  219104  20453605.9 < 2.2e-16 ***
JUN           1 147928  147928  13809261.0 < 2.2e-16 ***
JUL           1 444049  444049  41452487.1 < 2.2e-16 ***
AUG           1 250046  250046  23342089.4 < 2.2e-16 ***
SEP           1 145925  145925  13622230.8 < 2.2e-16 ***
OCT           1  63590   63590   5936156.1 < 2.2e-16 ***
NOV           1  28726   28726   2681561.7 < 2.2e-16 ***
DEC           1   7915    7915    738849.0 < 2.2e-16 ***
Residuals 102      1       0
---
Signif. codes:  0 '***' 0.001 '**' 0.01 '*' 0.05 '.' 0.1 ' ' 1
```

CONCLUSION

This chapter focused on various types of data like binary data, categorical data, and continuous data from different kinds of domains such as medical, agriculture, and socially relevant data. Regression is a kind of supervised machine learning algorithm which consists of various types. Linear regression is applied to child abuse data, and multiple linear regression is used on rainfall data. Autism contains binary data, so, logistic regression is applied to this data. All the models produced an adequate level and are in good health. In the future, other types of regression methods can be compared with these datasets to identify the opt model for the data.

CONSENT FOR PUBLICATION

Not applicable.

CONFLICT OF INTEREST

There is no conflict of interest declared.

ACKNOWLEDGEMENTS

Declared none.

REFERENCES

[1] *EMC, Data Science and Big Data Analytics.* Wiley, 2015.

[2] Chulka Sagar, *supervised machine learning,* 2016. https://www.geeksforgeeks.org/regression-classification-supervised-machine-learning/

[3] Sunil Ray, *Types of regression,* August 2015. https://www.analyticsvidhya.com/blog/2015/08/comprehensive-guide-regression/

[4] K. Himani, "Association Rule Mining:Algorithms used", In: *International Journal of Computer Science and Mobile Computing.* 2015, pp. 271-277.

[5] Geddam Jaishankar Rajkumar, "A Review Paper on Cricker Predictions Using Various Machine Learning Algorithms and comparisons among them", *International Journal for Research in Applied Science & Engineering Technology,* pp. 28-32.

[6] Nandhakumar Mahnedrapandian, "Needs, opportunities and challenges of big data analytics in the education field", *International Journal of Computer Engineering and Applications,* pp. 460-464, 2018.

[7] Abdo Ahmed Aqlan Hesham, and Ahmed Ajit Danti Shoieb, "Prediction of death based on target class variance and dimensionality reduction", *International journal of computer engineering and applications,* pp. 68-83.

[8] Herse Peter, and Omar Rokiah, "Quantification of dark adaptation dynamics in retinitis pigmentosa using non-linear regression analysis", *Clinical and Experimental Optometry,* pp. 386-389, 2004.

[9] I.D. Dinov, B. Heavner, M. Tang, G. Glusman, K. Chard, M. Darcy, R. Madduri, J. Pa, C. Spino, C. Kesselman, I. Foster, E.W. Deutsch, N.D. Price, J.D. Van Horn, J. Ames, K. Clark, L. Hood, B.M. Hampstead, W. Dauer, and A.W. Toga, "Predictive big data analytics: a study of parkinson's disease using large, complex, heterogeneous, incongruent, multi-source and incomplete observations", *PLoS One,* vol. 11, no. 8, 2016, e0157077.
[http://dx.doi.org/10.1371/journal.pone.0157077] [PMID: 27494614]

[10] Shaw Shobha Poudel Rajib, "The Relationships between Climate Variability and Crop Yield in Mountainous Environment", *MDPI,* pp. 1-19, March 2016.

[11] Mishra Debahuti, Hari Gour, and Subhadra Mishra Santra, "Applications of Machine Learning Techniques in Agricultural Crop Production: A Review Paper", *Indian Journal of Science and Technology,* pp. 1-14, 2016.

[12] Thomas Bilaliib Jianguo, and Ibn Musah Abdul-Aziz, "The nexus of weather extremes to agriculture production indexes and the future risk in ghana", *MDPI,* pp. 1-24.

[13] N. Kumar, and V.A.K. Vishal, "Efficient crop yield prediction using machine learning algorithms", *International research journal of engineering and technology,* vol. 5, no. 6, pp. 3151-3159, June 2018.

[14] S. Modi, Chourasiya, "Crop Prediction using Machine Learning," IOSR Journal of Engineering", *IOSR Journal of Engineering (IOSR JEN),* pp. 6-10, 2019.

[15] O. Buchade, Zingade, "Crop prediction system using machine learning", *International journal of advance engineering and research developement,* vol. 4, no. 5, pp. 1-6, 2017.

[16] N. Marlena, Kayleigh Novack, and Hyde K., *Applications of supervised machine learning in autism spectrum disorder research: a review.* Springer, 2019, pp. 1-19.

[17]   H.S. Alarifi, and G.S. Young, "Using multiple machine learning algorithms to predict autism in children", *International conference artificial intelligence,* 2018, pp. 464-469

[18]   M. Vinayakamurthy, V. Anuradha, and K. Vijayalakshmi, "Optimization of Machine Learning Techniques on Autism Spectrum Disorder with Swarm Intelligence Based Feature Selection", *International Journal of Recent Technology and Engineering,* pp. 6248-6251.

[19]   J. Matthew, and Maenner Scott H. Lee, "A comparison of machine learning algorithms for the surveillance of autism spectrum disorder", *PLoS One,* pp. 1-11, 2019.

[20]   Tawanchai Jirapramukpitak, "Child abuse and risky behaviors among youths", *Journal of the Medical Association of Thailand,* vol. 93, suppl. 7, pp. S160-S165, 2010.
[PMID: 21298839]

[21]   P. Gillingham, "Predictive risk modelling to prevent child maltreatment and other adverse outcomes for service users: inside the 'black box' of machine learning", *British Journal of Social Work,* vol. 46, no. 4, pp. 1044-1058, 2016.
[http://dx.doi.org/10.1093/bjsw/bcv031] [PMID: 27559213]

[22]   B.R.M. Divya, "Analyzing the Social Awareness in Autistic Children Trained through Multimedia Intervention Tool using Data Mining", *International Journal of Advanced Computer Science and Applications,* pp. 276-280.

[23]   Taboga Marco, "Generalised least square", 2018.

[24]   Daume Hal, *A Course in Machine learning,* 2012.

[25]   H. Krishna, and C.M.K. Ashok, "A fuzzy environment strategy for optimal agricultural land allocation in krishna delta", *International Research Journal of Computer Science,* vol. 5, no. 2, pp. 57-64, February 2018.

[26]   Faisal Zaman, Hideo hirose, and monira sumi, "A rainfall forecasting method using machine learning models and its application to the fukuoka city case", *International Journal Applied Mathematics and Computer Science,* vol. 22, no. 4, pp. 841-854, 2012.
[http://dx.doi.org/10.2478/v10006-012-0062-1]

[27]   T.A.O. Fulu, ZHANG Zhao, and SHI Wenjiao, "A review on statistical models for identifying climate contributions to crop yields", *Journal of Geographical Sciences,* vol. 23, no. 3, pp. 567-576, 2013.
[http://dx.doi.org/10.1007/s11442-013-1029-3]

[28]   K.A. Acharjya, "A Survey on Big daata analytics:Challenges, Open research Issues and Tools", In: *International of Computer science and Applications,* vol. 7, no. 2, pp. 511-518, 2016.

[29]   Y. Menaka, "A Survey on Crop Yield Prediction Models", *Indian Journal of Innovations and Developments,* vol. 5, no. 12, pp. 1-7, October 2016.

[30]   S.K.P. Chellammal, "An approach for prediction of crop yield using machine learning and big data techniques", *International journal of computer engineering and technology,* vol. 10, no. 3, pp. 110-118, June 2019.

[31]   Varun Khatri, "Application of Fuzzy logic in water irrigation system", *International Research Journal of Engineering and Technology,* pp. 3372-3375, April 2018.

[32]   Arunkumar Alkachoudhry, *Descriptive statistics.* Krishna prakasan media ltd: meerut, 2010.

[33]   Christos Evangelides, Christos Vrekos, and Christos Tzimopoulos, "Fuzzy Linear Regression of Rainfall-Atitude Relationship", *MDPI,* pp. 1-9, August 2018.

CHAPTER 2

Quantifying Players' Monopoly in a Cricket Team: An Application of Bootstrap Sampling

Bireshwar Bhattacharjee* and **Dibyojyoti Bhattacharjee**

Department of Statistics, Assam University, Silchar, Pin: 788011, Assam, India

Abstract: Cricket is a bat-and-ball game. It is played between two teams, each team consisting of 11 players. In limited-overs cricket, the teams play for a fixed number of overs, usually 50 or 20. At the end of the match, the team which scores the most number of runs in those limited-overs win the match. In this paper, taking the data from ICC Cricket World Cup 2019, an attempt is made to identify the type of competition that exists between the players, *i.e.*, batsmen and bowlers using the Herfindahl-Hirschman Index (HHI). This index is a statistical device used for estimating the degree of concentration in a particular market. A team is said to have the monopoly in scoring runs if the bulk of their scoring in the tournament is done by a few batsmen only while the other batsmen made an insignificant contribution with the bat. Likewise, a team is said to have bowler's monopoly, if the majority of wickets is taken by few bowlers of the team while the other bowlers could dismiss an insignificant number of opponent batsmen in the tournament. Applying bootstrap sampling, the teams are classified into three groups *viz.* monopoly, moderately competitive, and perfectly competitive. From the analysis, it is found that India, Australia, Bangladesh, and New Zealand are the teams where a monopoly exists, *i.e.*, most numbers of runs are scored by two or three batsmen. All other teams except Pakistan, *i.e.*, Afghanistan, West Indies, Sri Lanka, South Africa, England, are categorized as having perfect competition in the task of run-scoring. On the other hand, in the case of bowlers, Australia, Pakistan, Sri Lanka, and Bangladesh enjoys monopolistic nature in bowling. All other teams such as India, New Zealand, West Indies, Afghanistan, South Africa, and England are categorized as having Perfect Competition in the task of taking wickets. The study finds that out of the four semi-finalists, three of the teams enjoy a monopoly of batsmen, and three teams enjoy the perfect competition of bowlers. Thus, the work concludes that the monopoly of batsmen in a cricket team and perfect competition amongst bowlers have a role to play in the performance of teams in the tournament.

Keywords: Bootstrapping, Cricket, Cricket analytics, HHI, ICC cricket world cup, monopoly.

***Corresponding author Bireshwar Bhattacharjee:** Research Scholar, Department of Statistics, Assam University, Silchar, Pin-788011; Tel: +91-9531045037; Email: bireshwarbhattacharjee09@gmail.com.

INTRODUCTION

Cricket is a bat-and-ball game. The game is played in three different formats known as Test, ODI, and Twenty20 Cricket. Test cricket is the longest format, which runs to five days and two innings each team. Twenty20 cricket is the shortest form of cricket, with a maximum of twenty overs each side. It is played between two teams, each team consisting of 11 players. One-team bats (Team A, say) while the other team fields and bowls (Team B, say). Team A tries to score as many runs as possible against the bowling of Team B, who tries to dismiss the batsmen and hence minimizes the runs scored by Team A [1].

One Day International (ODI) is a form of limited-overs cricket, played between two teams with international status, in which each team faces a fixed number of overs, usually 50. The first ODI was played on 5th January 1971 between Australia and England at the Melbourne Cricket Ground. The One Day Cricket World Cup is generally held every four years. The first World Cup was organized in England in June 1975, which was won by West Indies. Australia lifted the trophy for a maximum number of five times, West Indies, and India lifted the trophy two times, and Pakistan, Sri Lanka, and England won it once [1].

In this paper, an attempt is made to identify the major contributory batsmen and bowler and their impact on their team's performance. This study is based on data from the ICC Cricket World Cup 2019 because after 1992, the type of format was first introduced where each team plays against every team once, and there would be 9 games for each team in the group stage since the participating teams are restricted to 10.

This paper is divided into seven subsections. The first subsection deals with the introduction; this is followed by the motivation of the study. Third section deals with the review of the literature. The objectives of the study are illustrated in the fourth subsection. Fifth subsection deals with methodology. Results and discussion are provided in the sixth subsection. The last section elaborates on the conclusions and directions for further research.

MOTIVATION OF THE STUDY

Cricket is a team game, where batting and bowling are the prime skills of the game. A better team shall have several good batsmen as well as superlative bowlers of different specialization who perform as and when demanded to make sure that their team wins against their opponent. But having few quality batsmen and bowlers in the team may not always reflect in the team performance as other team members may not perform efficiently, leading to excessive dependence of the team on few players. This may lead to a situation where the chance of victory of a team gets reduced. Thus, it has been felt to design a study to ascertain the type of competition that exists between the teams that participated in the Cricket World Cup of 2019 in terms of their batting and bowling performance.

REVIEW OF LITERATURE

In this section, we try to discuss some literature related to the quantification of the performance of batsmen, bowlers, and that of teams' performance. Though in cricket, there are several available measures to quantify the performance of players, like batting average and strike rate for batsmen, bowling average, bowling strike rate and economy rate for bowlers and percent of victory for teams; such measures always do not reveal the true level of performance of the cricketers and teams. Accordingly, different researchers have contributed significantly to defining several measures of performance analysis in cricket.

Kimber and Hansford [2] observed that the batting average while quantifying the batsman performance is ubiquitous in cricket, so they concluded that the traditional batting average depends on an unrealistic parametric assumption. They proposed a non-parametric approach based on runs scored for assessing batting performance. Some other works related to batting performance are conducted by Wood [3], Barr and Kantor [4] Maini, and Narayanan [5] Borooah and Mangan [6], *etc.*

Lemmer [7], while analyzing the performance of players, developed a current bowling performance measure (CBR), which is a joint measure that takes into account all the important measures of bowling performance. Its calculation requires the use of an easily programmable algorithm using only the bowler's career values of overs bowled, runs given, and wickets per innings played. Some other works related to bowling performance were performed Beaudoin and Swartz [8], Garber and Sharp [9], Dey *et al.*, [10] Bhattacharjee *et al.*, [11], *etc.*

Passi and Pandey [12] attempted to predict the performance of players- batsman and bowlers. This is done by classifying the number of runs scored, and the number of wickets dismissed in different ranges. They have used naive Bayes, random forest, multiclass SVM and decision tree classifiers to generate the prediction models for both the categories.

Owen and Owen [13] measured the degree of competitive balance among the teams to solve a common problem with competitive balance measures, *i.e.*, their sensitivity to the number of teams and the number of matches played by each team, known as season length. Their paper uses simulation methods to examine the effects of changes in season length on the distributions of several widely used variants of the Herfindahl-Hirschman index applied to wins in a season. Of the measures considered, a normalized measure, accounting for lower and upper bounds, and an adjusted measure perform best, although neither of them completely removes biases associated with different season lengths. Some other works related to team performance were performed Suleman and Saeed [14], Ghosh *et al.* [15], *etc.*

As cricket is a team game, so a better team needs to have several winning batsmen as well as bowlers who perform in a way to make sure that their team wins against the opponent in all situations. But having few superlative batsmen and bowlers may not always lead to the team's victory if the key players fail to perform. In such a situation, it is important to know whether in a team, all the players are performing or the team is heavily dependent on a few players only. Because very few works in the available literature have addressed the competitive balance of a team, this paper tries to quantify the extent of competition among the players of a cricket team in terms of the prime skills of the game *viz.* batting and bowling.

OBJECTIVES OF THE STUDY

The study is framed to attain the following two objectives:

- To determine the extent of competition amongst the batsmen of a cricket team.
- To determine the extent of competition amongst the bowlers of a cricket team.

METHODOLOGY

Ismail [16] studied the characteristics of market structures and methods of measuring the concentration as knowing the market structure and setting the level of competition is important for the policy to be followed for decision-makers. He observes that for determining market structures, several concentration criteria are commonly used. Some of them include concentration ratios, Herfindahl-Hirschman Index (HHI), Lorenz curve, Gini coefficient, Rosenbluth index, Entropy index, Linda index, Horwath index, Lerner index, and so on. The most widely used is the Herfindahl - Hirschman Index (HHI). The Herfindahl-Hirschman Index (HHI) is devised for estimating the degree of concentration in a particular market. It is calculated by squaring the market share of each firm competing in a market and then summing the resulting numbers. It can range from 0 to 10,000. [17] Here we used the HHI for determining the concentration or monopoly of batsmen's and bowlers within a cricket team based on their batting and bowling performances. The formula for calculating HHI is given by

$$\text{HHI(Batsman)} = \sum S_i^2 \tag{1}$$

Where,

$$S_i = \frac{\text{Total Runs scored by the } i^{th} \text{ batsman}}{\text{Total runs scored by the team in which the } i^{th} \text{ batsman played}} \times 100 \tag{2}$$

$$\text{HHI(Bowler)} = \sum T_j^2 \tag{3}$$

Where,

$$T_j = \frac{\text{Total wickets taken by the } j^{th} \text{ bowler}}{\text{Total wickets taken by the team in which the } j^{th} \text{ bowler played}} \times 100 \tag{4}$$

We compute HH Index for all the teams participated in the ICC Cricket World Cup 2019 for batting and bowling separately. On computing the same for all the ten teams, we bootstrap these 10000 times using the R package. *R* is a *programming* language and free software environment for statistical computing and graphics supported by the *R* foundation for statistical computing. The *R* language is widely used among statisticians and data miners for developing statistical software and data analysis. [18]. Bootstrapping is a technique of random sampling with replacement, *i.e.*, in bootstrapping, we can generate a large number of samples from a small sample. Bootstrapping allows assigning measures of accuracy, *i.e.*, in terms of bias, variance, *etc.* [19]. This technique allows estimation of the sampling distribution of almost any statistic using random sampling methods. Generally, it falls in the broader class of resampling methods [20]. Bootstrapping results are used to calculate the 50th and 75th percentile values of the HHI, *i.e.*, both for S_i and T_j separately. This shall be done to classify the teams into three categories, namely monopoly, moderate competition, and perfect competition. Here HH Index higher than 75th percentile is categorized as teams that are enjoying monopoly for batting and bowling. Teams belonging to this category are identified as those teams where two or three batsmen/bowlers have scored/taken the maximum number of runs/wickets, respectively. Similarly, teams with HHI who fall between 50th and 75th percentile are those teams where there is moderate competition amongst the batsmen/bowler. Likewise, a team with HHI less than 50th percentile are categorized as teams that have perfect competition, where there are several batsmen and bowler who contributed towards the team.

Data Source

The data for the analysis were extracted from the secondary sources. The website of ESPN dedicated to cricket data is espncricinfo.com, which is the data source for this work.

Data Collection Process

Data was collected from espncricinfo.com with the help of online software, Webharvy. It is a paid online web scraper used for extracting data from a website. The data of all the ODI matches played by various teams in the ICC World Cup 2019 as available in the Web Pages of espncricinfo.com were extracted to Microsoft Excel files using the above mentioned Web Scrapper.

RESULTS AND DISCUSSION

The percentage of runs scored by two leading batsmen for each of the ten teams and the percentage of wickets taken by two leading bowlers for each of the ten teams were collected from the source mentioned above. Also, the sum of the percentage of runs scored and the number of wickets were calculated as illustrated in Tables **1** and **2**.

The top three contributory batsmen for their respective teams are Kane Williamson (28.57 percent), Sakib-Ul Hasan (28.25 percent) and Rohit Sharma (26.64 percent). It may also be noted that K Williamson and R Taylor of New Zealand and S Al Hasan and M Rahim of Bangladesh in unison scored more than 45 percent of the total runs for their respective teams (Table **1**).

Table 1. Major contributory (top two) batsmen of the teams in terms of runs scored (*Authors computation based on data collected from espncricinfo.com*).

Country	Rank	Batsman	Percentage of Runs Scored	Combined %
New Zealand	1	K Williamson	28.57	45.96
	2	R Taylor	17.30	
Bangladesh	1	S Al Hasan	28.25	45.35
	2	M Rahim	17.10	
India	1	RG Sharma	26.64	44.85
	2	V Kohli	18.21	

(Table 1) cont.....

Australia	1	D Warner	23.47	41.85
	2	A Finch	18.38	
Pakistan	1	B Azam	24.51	40.28
	2	Imam Ul Haq	15.77	
South Africa	1	F Du Plessis	21.05	37.97
	2	V D Dussen	16.92	
England	1	J Root	18.18	35.57
	2	J Bairstow	17.39	
West Indies	1	N Pooran	20.01	34.95
	2	S Hope	14.94	
Sri Lanka	1	K Perera	18.16	34.39
	2	A Mathews	16.23	
Afghanistan	1	R Shah	14.80	28.2
	2	N Zadran	13.40	

Again, the top three contributory bowlers are Mitchell Starc (43.55 percent), Lasith Malinga (38.23 percent), and Mustafizur Rahman (33.90 percent). It may also be noted that M Starc and R Taylor of Australia; M Rahman and M Saifuddin of Bangladesh and L Malinga and I Udana in unison dismissed more than 55 percent of the total batsmen dismissed by the bowlers of their respective teams (Table **2**).

Table 2. Major contributory (top two) bowlers of the teams in terms of wickets taken (*Authors computation based on data collected from espncricinfo.com*).

Country	Rank	Bowler	Percentage of Runs Scored	Combined %
Australia	1	M Starc	43.55	66.13
	2	R Taylor	22.58	
Bangladesh	1	M Rahman	33.90	55.93
	2	M Saifuddin	22.03	
Sri Lanka	1	L Malinga	38.23	55.88
	2	I Udana	17.65	
Pakistan	1	M Amir	28.33	55.00
	2	S Afridi	26.67	
New Zealand	1	L Ferguson	25.61	46.34
	2	T Boult	20.73	
India	1	J Bumrah	24.32	43.23
	2	M Shami	18.91	
Afghanistan	1	M Nabi	21.28	40.43
	2	G Naib	19.15	
England	1	J Archer	22.22	40.00
	2	C Woakes	17.78	
South Africa	1	C Morris	21.67	40.00
	2	I Tahir	18.33	
West Indies	1	S Cottral	22.22	38.89
	2	O Thomas	16.67	

Now we can distinguish teams into three categories, such as Monopoly, Perfect competition, and Moderate Competition based on their HHI value. Here more than 75 percent value is taken as a monopoly, moderate competition is taken between 50 to 75 percent, and less than 50 percent is taken as perfect competition. The criteria for the classification of the type of competition for both batsmen and bowlers of the teams are discussed in Tables **3** and **4** below.

Table 3: Criteria for classifying teams for the batsman.

	In Percentage	Value of HHI (Batsman)
Monopoly	Greater than 75%	Greater than 1408
Moderate competition	50% to 75%	1374 to 1408
Perfect competition	Less than 50%	Less than 1374

Table 4: Criteria for classifying teams for bowlers.

	In Percentage	Value of HHI (Bowler)
Monopoly	Greater than 75%	Greater than 2019
Moderate competition	50% to 75%	1926 to 2018
Perfect competition	Less than 50%	Less than 1925

Now using the criterion discussed above (Table **3**), the HHI value greater than 1408 is categorized as a monopoly and indicated by white colour, moderate competition lies between 1374 to 1408, which is indicated by grey color, and perfect competition falls below 1374 and is indicated by black colour (Fig. **1**).

Based on the criterion discussed above (Table **4**), the HHI value greater than 2019 is categorized as a monopoly and indicated by white colour, moderate competition lies between 1926 to 2018, and lastly, perfect competition is taken as below 1925. It is indicated by the black color (Fig. **2**).

The HHI for India, New Zealand, Bangladesh, and Australia are quite high; thus, based on classification criteria as discussed in Table **3**, these teams show a monopoly feature among batsman, *i.e.*, these teams are heavily dependent on few players to perform so far as batting is concerned. Pakistan is the only team where the competition amongst batsmen is moderate. All other teams, *i.e.*, Afghanistan, West Indies, Sri Lanka, South Africa, and England, are categorized as having perfect competition as their HHIs are low, *i.e.*, the batsmen of these teams share the task of run-scoring (Fig. **1**).

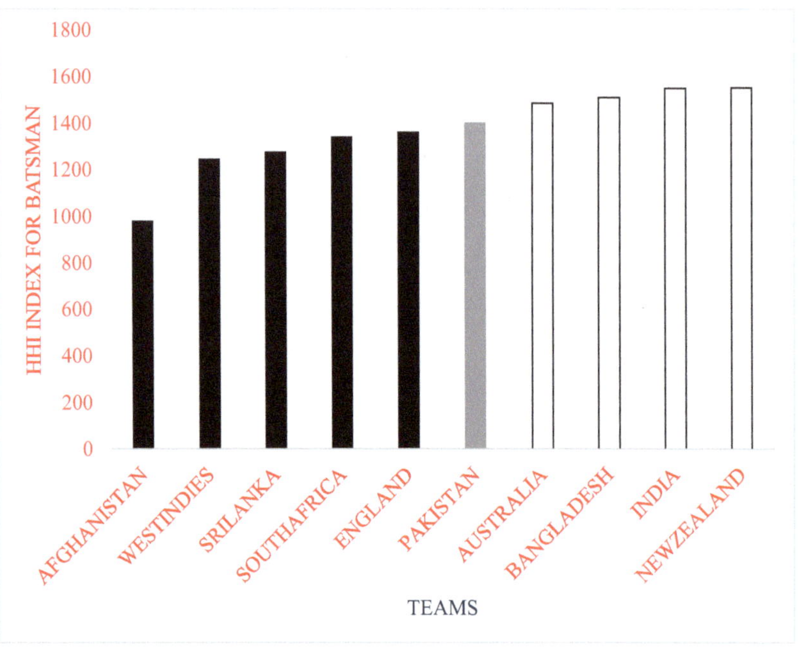

Fig. 1. HHI for a batsman of different cricket teams based on data from World Cup 2019 (Authors Computation).

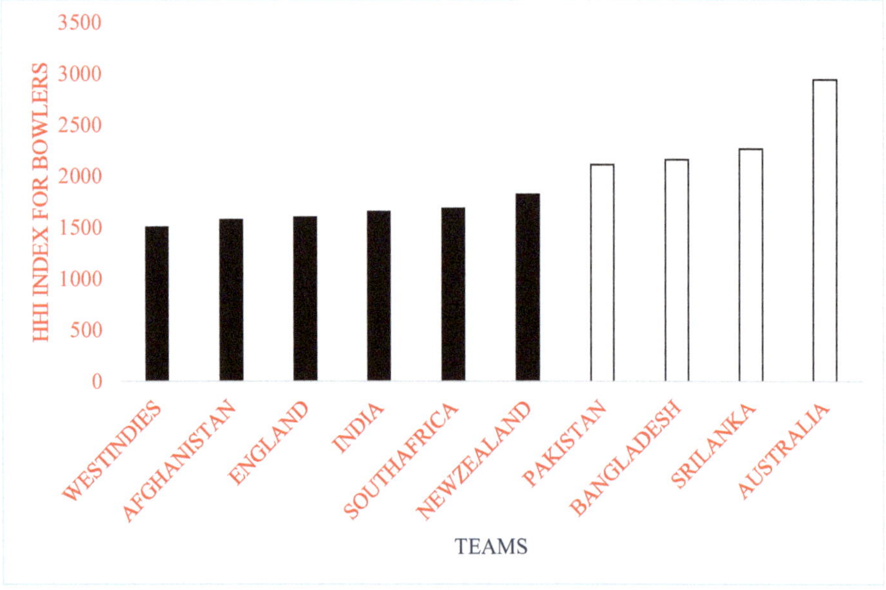

Fig. 2. HHI for bowlers of different cricket teams based on data from World Cup 2019.

Again the HHI for Australia, Pakistan, Sri Lanka, and Bangladesh is quite high; thus, based on classification criteria as discussed in Table **4**, these teams show a Monopoly feature among bowlers, *i.e.*, these teams are heavily dependent on few players to perform so far as bowling is concerned. The teams *viz* India, New Zealand, West Indies, Afghanistan, South Africa, and England, are categorized to having perfect competition in bowling as their HHIs are low, *i.e.*, the bowler of these teams share the task of taking wickets (Fig. **2**).

CONCLUSION

Thus, from this analysis, it is clear that among the batsmen of India, Australia, Bangladesh, and New Zealand, there is a monopoly, *i.e.*, the bulk of the runs scored by these teams are concentrated amongst two to three players. Moderate competition exists between the batsmen of Pakistan, and perfect competition exists amongst batsmen in the teams of Afghanistan, West Indies, Sri Lanka, England, and South Africa. Whereas in the case of bowlers, Australia, Sri Lanka, Pakistan, and Bangladesh have shown monopolistic nature and all other teams, including India, South Africa, England, Afghanistan, New Zealand, and West Indies have perfect competition among their bowlers. It may be noted that the top four teams of the ICC Cricket World Cup, 2019 were England, New Zealand, India, and Australia. Out of these four teams, three of them had a monopolistic nature in terms of run-scoring. However, in case of bowling, three of the first four teams, *viz.* England, India, and New Zealand had a perfect competition amongst the bowlers of the respective teams. Thus, in the said tournament having a few specialist batsmen in the team who can score the bulk of the runs and having several bowlers with the ability to take wickets has been proved to be the key characteristics for success. However, to generalize these findings, a similar exercise needs to be performed for several such tournaments involving multiple numbers of teams to compare the performance of teams concerning their HHI value.

CONSENT FOR PUBLICATION

Not applicable.

CONFLICT OF INTEREST

There is no conflict of interest declared.

ACKNOWLEDGEMENTS

Declared none.

REFERENCES

[1]   "History of Cricket," Apr. 1, 2019. [Online]. Available: www.wikipedia.org. [Accessed: 5th Mar. 2020].

[2]   A.C. Kimber, and A.R. Hansford, "A Statistical Analysis of Batting in Cricket", *J. R. Stat. Soc. Ser. A Stat. Soc.,* vol. 156, pp. 443-455, 1993.
[http://dx.doi.org/10.2307/2983068]

[3]   G. Wood, "Cricket Scores and Geometric Progression", *Journal of Royal Statistical Society,* vol. 108, pp. 12-22, 1945.

[4]   G. Barr, and B. Kantor, "A Criterion for Comparing and Selecting Batsman in Limited Overs Cricket", *Journal of the Operational Research Society,* vol. 55, no. 12, pp. 1266-1274, 2004.
[http://dx.doi.org/10.1057/palgrave.jors.2601800]

[5]   S. Maini, and S. Narayanan, "The Flaw in Batting Averages," The Actuaries, pp. 30-31, 2007.

[6]   V.K. Borooah, and J. Mangan, "The "Bradman Class": An exploration of Some Issues in the Evaluation of Batsman for Test Matches", *Journal of Quantitative Analysis in Sports,* vol. 6, no. 3, pp. 1877-2006, 2010.
[http://dx.doi.org/10.2202/1559-0410.1201]

[7]   H.H. Lemmer, "A Measure of the Current Bowling Performance in Cricket", *South African Journal for Research in Sport, Physical Education and Recreation,* vol. 28, no. 2, 2006.
[http://dx.doi.org/10.4314/sajrs.v28i2.25946]

[8]   D. Beaudin, and T. Swartz, "The Best Batsman and Bowlers in One Day Cricket", *South African Statistical Journal,* vol. 37, pp. 203-222, 2003.

[9]   H. Gerber, and G.D. Sharp, "Selecting a limited Overs Cricket Squad Using an Integer Programming Model", *South African Journal for Research in Sport, physical Education and Recreation,* vol. 28, no. 2, pp. 81-90, 2006.
[http://dx.doi.org/10.4314/sajrs.v28i2.25945]

[10]   P. Dey, D.N. Ghosh, and A.C. Mondal, "A MCDM Approach for Evaluating Bowlers Performance in IPL", *Journal of Emerging Trends in Computing and Information Sciences,* vol. 2, no. 11, pp. 563-573, 2011.

[11]   H. Saikia, D. Bhattacharjee, and H.H. Lemmer, "Predicting the Performance of Bowlers in IPL: An Application of Artificial Neural Network", *International Journal of Performance Analysis in Sport,* vol. 12, pp. 75-89, 2012. a
[http://dx.doi.org/10.1080/24748668.2012.11868584]

[12]   K. Passi, and N. Pandey, "Increased Prediction Accuracy in the Game of Cricket using Machine Learning", *International Journal of Data Mining and Knowledge Management Process,* vol. 8, no. 2, 2018.
[http://dx.doi.org/10.5121/ijdkp.2018.8203]

[13]   P.D. Owen, and C.A. Owen, "Simulation Evidence on Herfindahl-Hirschman Indices as Measures of Competitive Balance", *Economics Discussion Paper No. 1715, University of Otago:Otago,* 2017.

[14]   M. Suleman, and M. Saeed, "Option on Human Performance: A Case Study of Indian Premier League", *SSRN,* Sept. 2009.

[15]   P. Dey, D.N. Ghosh, and A. Mondal, "IPL Team Performance analysis: A Multi Criteria Group Decision Approach in Fuzzy Environment", *International Journal of Information Technology and Computer Science,* vol. 08, pp. 8-15, 2015.
[http://dx.doi.org/10.5815/ijitcs.2015.08.02]

[16]   I. Ukav, "Market Structures and Concentrsation Measuring Techniques", *Asian Journal of Agricultural Extension, Economics and Sociology,* vol. 19, no. 4, pp. 1-16, 2017.

[http://dx.doi.org/10.9734/AJAEES/2017/36066]

[17]   "Background HH Index", 2019. [Online]. Available: www.investopedia.com. [Accessed: 15th Feb. 2020].

[18]   "R programming language", Jun. 1, 2018. [Online]. Available: https://en.wikipedia.org/wiki/ R_(programming_language) [Accessed: 3rd Feb. 2020].

[19]   Y. Lorna, "An Introduction to the Bootstrap Method", 2019. [Online]. Available: https://towardsdatascience.com/an-introduction-to-the-bootstrap-method-58bcb51b4d60.

[20]   F.M. Clara, and M. Novoa, "Bootstrap methods for analyzing time studies and input data for simulations", *International Journal of Productivity and Performance Management,* vol. 58, no. 5, pp. 460-479, 2019.

CHAPTER 3

On Mean Estimation Using a Generalized Class of Chain Type Estimator under Successive Sampling

Shashi Bhushan[1], Nishi Rastogi[2] and Shailja Pandey[1,*]

[1] Department of Mathematics and Statistics, Dr. Shakuntala Misra National Rehabilitation University, Lucknow, India
[2] Department of Statistics, National P.G. College, Lucknow, India

Abstract: The present paper comprises a significantly generalized class of chain type estimators to estimate the population means on the current occasion under the framework of successive sampling based on auxiliary information on both the occasion. The proposed generalized class constitutes two renowned chain type classes proposed by Singh and Vishwakarma [1, 2]. As its particular case, an improvement over their notion, with some eased regularity conditions, is proposed by us, which consists of chain type regression estimators additionally to chain type ratio estimators. The construction of the proposed class is fruitful in the sense of constructing the chain type classes of estimators in the realm of successive sampling. In terms of efficiency, we provide a comparative study of the proposed class oversample mean estimator, Cochran's estimator [3], Sukhatme *et al.* estimator [4] and Singh's estimator [5]. A numerical illustration is demonstrated in support of the proposed class.

Keywords: Auxiliary information, Generalized class of chain type estimator, Optimum replacement policy, Successive sampling.

INTRODUCTION

In practice, the structure of the population under study remains stable over the given period while in applied sciences, sociology, and economic researches, this composition of the population changes over the same span. In such situations, successive sampling over different points of time equips a useful statistical tool for estimating the reliable numerical estimates. A common phenomenon to improve the precision on the most recent occasion in successive sampling, we use the estimates obtained on earlier occasions. This approach of drawing sample on the successive occasion was set up by Jessen [6] in the analysis of an agricultural survey

**Corresponding author Shailja Pandey:* Department of Mathematics and Statistics, Dr. Shakuntala Misra National Rehabilitation University, Lucknow, India; Tel: +91-7271020995; E-mail: writetoshailja6693@gmail.com.

data. For estimating more precise estimates on the current occasion, he availed all the information available. A similar notion of estimation was lengthened by Patterson [7], Rao, and Graham [8], Sen [9], Gupta [10], Das [11], Chaturvedi and Tripathi [12], and among others. It has been observed that the auxiliary information in sample surveys plays a supporting key role in enhancing the efficiency of the estimators. Therefore, besides information available on the previous occasion, continuous sampling yields more efficient estimates in the existence of auxiliary information. In many circumstances, for the survey purpose, this information may be available readily or made available by adjusting the small part of the available cost. The use of such type of information on two occasions successive sampling for estimating the population mean has been made by Feng and Zou [13], Biradar and Singh [14], Singh and Singh [15], Singh [16], Singh [5] and Singh and Vishwakarma [1, 2].

An earnest effort has been made in the sense of chain aspect, which covered both chain type ratio and regression estimators, in presenting a more general approach than Singh and Vishwakarma [2]. The presented approach also enhances the idea of constructing a chain type class of estimators in successive sampling. Now, Singh and Vishwakarma's [1, 2] classes are the subclasses of the proposed generalized class in *Subsection 2.2*. The coverage of numerous estimators under the proposed generalized chain class is given in *Subsections 2.3* and *2.4*. The minimum mean square error (MSE) of the proposed class is discussed in *Section 3*. Analytical and empirical studies are included in *Sections 4* and *5,* respectively.

SAMPLING METHODOLOGY

Sample Structure and Notations

Let us consider a framework of population size N , which we have to sample over two successive occasions with x and y as the characteristics understudy on the first and second occasions, respectively. Further, we have considered z the auxiliary information readily obtainable on the first and second occasions. For simplicity, we provide results for a large population with a sample of size n on both occasions. By adopting simple random sampling without replacement on the first occasion, we draw a sample of size n . Further, on the second occasion, we retain a subsample of size $m = n(1 - \theta)$ from the sampled units of the first occasion and $u = n\theta$ units from the units that were not selected on the first occasion, *i.e.,* $N - n$ units. We shall use the following notations for further use:

$\overline{X}$ and $\overline{Y}$: The population means of study variable x and y on the first and second occasions, respectively.

$\overline{Z}$: The population mean of the readily available auxiliary variable on the first and second occasions both.

$\overline{y}_u$ and $\overline{y}_m$: The sample means of the study variable on the second occasion based on sample sizes u and m respectively.

$\overline{x}_m$ and $\overline{x}_n$: The sample means of the study variable on the first occasion based on sample sizes shown in subscripts m and n, respectively.

ρ_{yx}, ρ_{yz} and ρ_{xz}: These are the correlation coefficients between the variables written in their subscript.

S_y^2, S_x^2 and S_z^2: These are sampling variance of the variables displayed in subscript.

θ: The unmatched proportion of the sample units

FORMULATION OF THE PROPOSED GENERALIZED CLASS

Following Singh and Vishwakarma [1, 2] and Bhushan and Rastogi [17], we formulate a new generalized class of chain type estimator to estimate the population mean on the current occasion as follows:

$$T = \phi T_H' + (1-\phi)T_G' \tag{1}$$

here, T_H' is the class of estimators for unmatched proportion as considered by Singh and Vishwakarma [1]

$$T_H' = H'(\overline{y}_u, \overline{z}_u) \tag{2}$$

Where H' is such that $H'(\overline{Y}, \overline{Z}) = \overline{Y}$ also satisfies the following conditions in a bounded, closed convex subset with the real space $\Re$ of dimension two:

(i) H' the function is continuous and bounded in $\Re$.

(ii) The partial derivatives of the first and second-order of H' existence. These are also continuous and bounded in real space $\Re$.

(iii) The partial derivative of H' with respect to $\overline{y}_u$ is 1, *i.e.* $\left. \dfrac{\partial H'}{\partial \overline{y}_u} \right]_{\overline{Y},\overline{Z}} = H'_1 = 1$.

Let $\overline{y}_m$, $\overline{x}_m$, $\overline{x}_n$, $\overline{z}_m$ and $\overline{z}_n$ are the unbiased estimators of $\overline{Y}$, $\overline{X}$, $\overline{X}$, $\overline{Z}$ and $\overline{Z}$ based on units given in subscripts of sample means, respectively. Let $(\overline{y}_m, \overline{x}_m, \overline{x}_n, \overline{z}_m, \overline{z}_n)$ assume values in a closed convex subset of five-dimensional real space enclosing the point $(\overline{Y}, \overline{X}, \overline{X}, \overline{Z}, \overline{Z})$.

By following Bhushan and Rastogi [17], we have proposed T'_G as the class of chain type estimators for the matched portion

$$T'_G = G'(\overline{y}_m, \overline{x}_m, \overline{x}_n, \overline{z}_m, \overline{z}_n) \tag{3}$$

where the function G' is such that $G'(\overline{Y}, \overline{X}, \overline{X}, \overline{Z}, \overline{Z}) = \overline{Y}$ and also satisfies the following conditions:

(i) G' the function is continuous and bounded in $\Re$.

(ii) The partial derivatives of the order first and second of G' existence. These are also continuous and bounded in real space $\Re$.

(iii) The partial derivative of G' with respect to $\overline{y}_m, \overline{x}_m, \overline{x}_n, \overline{z}_m$ and $\overline{z}_n$ are denoted as G'_1, G'_2, G'_3, G'_4 and G'_5 at the point $(\overline{Y}, \overline{X}, \overline{X}, \overline{Z}, \overline{Z}) = \overline{Y} = P$ and linked by the following regularity conditions:

$$G'_1 = 1$$

$$G'_2 = -G'_3$$

$$G'_4 = -(G'_5 + \gamma) \; ; \; \gamma \text{ is scalar, which is obtained under optimization of mean square error.}$$

IMPORTANT SPECIAL CASES OF CASES OF CLASS OF ESTIMATOR FOR UNMATCHED PROPORTION

We consider the same generalized class of unmatched proportion as considered by Singh and Vishwakarma [1]. This class for unmatched covers numerous estimators such as

$$\bar{y}_{u1} = \bar{y}_u \left(\frac{\bar{Z}}{\bar{z}_u} \right)$$

$$\bar{y}_{u2} = \bar{y}_u \left(\frac{\bar{z}_u}{\bar{Z}} \right)$$

$$\bar{y}_{u3} = \bar{y}_u \left(\frac{\bar{z}_u}{\bar{Z}} \right)^{\alpha_1}$$

$$\bar{y}_{u5} = \bar{y}_u \bar{Z} \Big/ \left[\bar{Z} + \alpha_3 (\bar{z}_u - \bar{Z}) \right]$$

$$\bar{y}_{u6} = \alpha_4 \bar{y}_u \left(\frac{\bar{Z}}{\bar{z}_u} \right) + (1 - \alpha_4) \left(\frac{\bar{Z}}{\bar{z}_u} \right)^2$$

$$\bar{y}_{u7} = \alpha_5 \bar{y}_u \left(\frac{\bar{z}_u}{\bar{Z}} \right) + (1 - \alpha_5) \left(\frac{\bar{z}_u}{\bar{Z}} \right)^2$$

$$\bar{y}_{u8} = \alpha_6 \bar{y}_u + (1 - \alpha_6) \left(\frac{\bar{Z}}{\bar{z}_u} \right)$$

$$\bar{y}_{u9} = \alpha_7 \bar{y}_u + (1 - \alpha_7) \left(\frac{\bar{z}_u}{\bar{Z}} \right)$$

$$\bar{y}_{u10} = \bar{y}_u \left[\alpha_8 \bar{Z} + (1 - \alpha_8) \bar{z}_u \right] \Big/ \left[\alpha_8 \bar{z}_u + (1 - \alpha_8) \bar{Z} \right]$$

Where the optimum value of constants is obtained from the optimum value of H_2' as mentioned in expression (7).

IMPORTANT SPECIAL CASES OF CLASS OF ESTIMATOR FOR MATCHED PROPORTION

The proposed generalized class for matched proportion contains Singh and Vishwakarma [1, 2] classes of the estimators. Therefore, the proposed class contains all those estimators which are undertaken by them. By following Chand [18], Srivastava [19] and Sahoo and Sahoo [20], a generalized class of estimators is suggested by Singh, and Vishwakarma [1] for matched proportion is

$$T_{jm} = J\left(\overline{y}_m, \overline{x}_m, \overline{x}_n, \overline{z}_n\right) \tag{4}$$

They discussed the following estimator, which covered the above-proposed class Chain type estimator which is taken by Singh [5], by following Chand's technique [18].

$$t_{jm1} = \overline{y}_m \left(\frac{\overline{x}_n}{\overline{x}_m}\right)\left(\frac{\overline{Z}}{\overline{z}_n}\right)$$

By motivation of Chand [18], one can also suggest the following estimator

$$t_{jm2} = \overline{y}_m \left(\frac{\overline{x}_n}{\overline{x}_m}\right)\left(\frac{\overline{z}_n}{\overline{Z}}\right)$$

$$t_{jm3} = \overline{y}_m \left(\frac{\overline{x}_m}{\overline{x}_n}\right)\left(\frac{\overline{Z}}{\overline{z}_n}\right)$$

Kiregyera [21, 22] motivation define:

Chain ratio to regression estimator

$$t_{jm4} = \left(\frac{\overline{y}_m}{\overline{x}_m}\right)\left[\overline{x}_n + \beta_{xz}\left(\overline{Z} - \overline{z}_n\right)\right]$$

The ratio in regression estimator

$$t_{jm5} = \overline{y}_m + \beta_{yx}\left[\left(\frac{\overline{x}_n}{\overline{z}_n}\right)\overline{\overline{Z}} - \overline{x}_m\right]$$

Regression in regression estimator

$$t_{jm6} = \overline{y}_m + \beta_{yx}\left[(\overline{x}_n - \overline{x}_m) + \beta_{xz}(\overline{\overline{Z}} - \overline{z}_n)\right]$$

Following Sahoo and Sahoo [20], one can define

$$t_{jm7} = (\overline{y}_m\overline{x}_m)\Big/\left[\overline{x}_n + \beta_{xz}(\overline{\overline{Z}} - \overline{z}_n)\right]$$

$$t_{jm8} = \overline{y}_m + \beta_{yx}\left[\overline{x}_n\left(\frac{\overline{z}_n}{\overline{\overline{Z}}}\right) - \overline{x}_m\right]$$

Further, Singh and Vishwakarma [2] proposed a chain type generalized sampling strategy for ratio estimators with the same sampling scheme as discussed in Singh and Vishwakarma [1] as follows

$$T_{Hm} = H(\overline{y}_m, a, b, c) \tag{5}$$

where, $a = \dfrac{\overline{x}_m}{\overline{x}_n}$, $b = \dfrac{\overline{z}_m}{\overline{z}_n}$, $c = \dfrac{\overline{z}_n}{\overline{\overline{Z}}}$

They discussed the following members who are covered in this class

$$t_{hm1} = \overline{y}_m a^{\delta_1} b^{\delta_2} c^{\delta_3}$$

$$t_{hm2} = \overline{y}_m(2 - a^{\delta_1} b^{\delta_2} c^{\delta_3})$$

$$t_{hm3} = \overline{y}_m + \delta_1(a-1) + \delta_2(b-1) + \delta_3(c-1)$$

$$t_{hm4} = \overline{y}_m\left[1 + \delta_4(a^{\delta_1} b^{\delta_2} c^{\delta_3} - 1)\right]^{-1}$$

$$t_{hm5} = \overline{y}_m\left[1 + \delta_1(a-1) + \delta_2(b-1) + \delta_3(c-1)\right]^{-1}$$

$$t_{hm6} = \bar{y}_m \left[1 - \delta_1(a-1) - \delta_2(b-1) - \delta_3(c-1) \right]$$

Where δ_1 δ_2, δ_3 and δ_4 the constants optimized by utilizing the different differential constants obtained in the expression of MSE of the estimator T_{Hm}.

Both the above important chain type classes of matched units and the estimators, which are part of both the families, are now members of the suggested chain type class.

MEAN SQUARE ERROR OF THE PROPOSED GENERALIZED CLASS

Convexity of a function plays an essential role in its optimization. Therefore, convex linearity of the proposed class provides optimum variance, which we obtained as follows

The Variance of the Units of the Unmatched Class T_H'.

For a large population, Singh and Vishwakarma [1] obtained the variance of T_H' up to the first degree of approximation, under the conditions discussed in the above section for T_H', as

$$V(T_{H'}) = f_u \left[S_y^2 + S_z^2 H_2'^2 + 2S_y S_z H_2' \rho_{yz} \right] \tag{6}$$

where, $H_2' = \dfrac{\partial H'}{\partial z_u} \bigg]_{(\bar{Y},\bar{Z})}$.

So, the variance of $T_{H'}$ is minimized H_2' as $H_2' = -\rho_{yz} \dfrac{S_y}{S_z}$ $\tag{7}$

Thus, the minimum variance T_H' is obtained as

$$\min.V(T_{H'}) = f_u S_y^2 (1 - \rho_{yz}^2) \tag{8}$$

The Variance of T'_G, the Class of Estimators for Matched Units.

Bhushan and Rastogi [17] obtained the variance of T'_G for large population size, under the two-phase sampling structure, up to the first-order degree of approximation, by imposing the conditions.

$$M(T_{G'}) = f_m S_y^2 + (f_m - f_n)(S_x^2 G_2'^2 + S_z^2 G_4'^2 + 2S_y S_x G_2' \rho_{yx} + 2S_y S_z G_4' \rho_{yz} + 2S_y S_z G_2' G_4' \rho_{xz})$$

$$+ f_n(\gamma^2 S_z^2 - 2S_y S_z \gamma \rho_{yz}) \tag{9}$$

Where, $G_2' = \dfrac{\partial G'}{\partial \overline{x}_m}\bigg]_P$, $G_3' = \dfrac{\partial G'}{\partial \overline{x}_n}\bigg]_P$, $G_4' = \dfrac{\partial G'}{\partial \overline{z}_m}\bigg]_P$ and $G_5' = \dfrac{\partial G'}{\partial \overline{z}_n}\bigg]_P$.

The variance $T_{G'}$ is minimized for G_2', G' and γ. The optimum values of these are obtained as follows:

$$G_2' = \frac{\rho_{xz}\rho_{yz} - \rho_{yx}}{1 - \rho_{xz}^2} \frac{S_y}{S_x}$$

$$G_4' = \frac{\rho_{xz}\rho_{yx} - \rho_{yz}}{1 - \rho_{xz}^2} \frac{S_y}{S_x}$$

and $\gamma = \rho_{yz} \dfrac{S_y}{S_z}$

Thus, the minimum variance of $T_{G'}$ is obtained by putting the above optimum values in expression (9) as follows

$$\min M(T_{G'}) = \left[f_m - (f_m - f_n)\rho_{y.xz}^2 - f_n \rho_{yz}^2 \right] S_y^2 \tag{10}$$

Theorem 1. The mean square error of the proposed class of chain type estimator is

$$\min. M(T) = \frac{1}{n} \frac{(1 - \rho_{yz}^2)\left[(1 - \rho_{yz}^2) + \theta(\rho_{yz}^2 - \rho_{y.xz}^2) \right]}{\left[(1 - \rho_{yz}^2) + \theta^2(\rho_{yz}^2 - \rho_{y.xz}^2) \right]} S_y^2 \tag{11}$$

where θ is optimized under the optimum replacement policy.

Proof. Since, in our sample structure, we have drawn u units from $(N-n)$ units. Therefore, there is no correlation between matched and unmatched units. The MSE of the generalized class T can be written as

$$M(T) = \phi^2 M(T_{H'}) + (1-\phi)^2 M(T_{G'}) \tag{12}$$

Now, we get the value of optimum ϕ by minimizing the above expression concerning ϕ, we have

$$\phi = \frac{M(T_{G'})}{M(T_{G'}) + M(T_{H'})}$$

On putting the above optimum value of ϕ in expression (12), we have

$$\min.M(T) = \frac{M(T_{G'})M(T_{H'})}{M(T_{G'}) + M(T_{H'})} \tag{13}$$

On putting Expressions (10) and (8) in the above expression, we obtained the resultant MSE, as mentioned in (11).

ANALYTICAL STUDY

We here provide an analytical approach to motivate our proposed class and its idea of construction. The inclusion of an optimum replacement policy provides optimum unmatched units, and the effectiveness of the proposed class is seen under efficiency comparison.

OPTIMAL REPLACEMENT POLICY

The knowledge of the fraction of unmatched and matched units is a key part of any sampling strategy under successive sampling. The optimization of unmatched (or matched) units is done by minimizing the expression of (11) with respect to the unmatched proportion θ, and we obtained the optimum value of θ as

$$\theta_{opt} = \frac{1 \pm \sqrt{1 - \rho_{yx.z}^2}}{\rho_{yx.z}^2} = \theta_0$$

Where, $\rho_{yx.z} = \dfrac{(\rho_{yx} - \rho_{yz}\rho_{xz})}{\sqrt{1-\rho_{xz}^2}\sqrt{1-\rho_{yz}^2}}$. Since. $\rho_{yx.z}^2$ lies between 0 & 1, and so

$(1-\rho_{yx.z}^2)$ is always positive. Therefore, the admissible value of θ_{opt} which lies between 0 & 1 is

$$\theta_{opt} = \frac{1-\sqrt{1-\rho_{yx.z}^2}}{\rho_{yx.z}^2} = \theta_0$$

Thus, the expression (11) reduces by utilizing θ_0 is

$$\min M^*(T) = \frac{1}{2n}\left(1+\sqrt{1-\rho_{yx.z}^2}\right)S_y^2 \tag{14}$$

EFFICIENCY COMPARISON

We assess the proposed generalized class of chain type estimator for its validation in the following way

1. **Natural mean per unit estimator.** The efficiency of the proposed class T with respect to the natural mean per unit estimator is given by

$$E_1 = \frac{V(\overline{y}_n)}{\min.M^*(T)}*100 = \frac{2}{\left(1+\sqrt{1-\rho_{yx.z}^2}\right)}*100$$

where, $V(\overline{y}_n) = f_n\overline{Y}^2 C_y^2$

2. **Cochran's estimator [3].** Cochran has suggested an estimator with no auxiliary information as follows

$$\overline{y}_\psi = \psi\overline{y}_u + (1-\psi)(\overline{y}_m + b(\overline{x}_n - \overline{x}_m))$$

Thus, the minimum variance of $\overline{y}_\psi$ is

$$V\left(\overline{y}_{\psi}\right)_{opt} = \frac{1}{2n}\left(1+\sqrt{1-\rho_{yx}^2}\right)S_y^2$$

with the optimum unmatched proportion

$$\theta_{opt} = \frac{1}{1+\sqrt{1-\rho_{yx}^2}} = \theta_1$$

3. **Sukhatme *et al.* estimator [4].** By taking difference estimator in the matched portion, Sukhatme considered the following estimator

$$\overline{y}_{\psi}^{*} = \psi^{*}\overline{y}_u + (1-\psi^{*})(\overline{y}_m + \beta(\overline{x}_n - \overline{x}_m))$$

The variance of the above estimator is the same, as obtained for Cochran's estimator [3].

$$V\left(\overline{y}_{\psi}^{*}\right)_{opt} = V\left(\overline{y}_{\psi}\right)_{opt} = \frac{1}{2n}\left(1+\sqrt{1-\rho_{yx}^2}\right)S_y^2$$

So, the efficiency concerning both the estimators $\overline{y}_{\psi}$ or $\overline{y}_{\psi}^{*}$ is the same. Thus,

$$E_2 = \frac{V(\overline{y}_{\psi} \ or \ \overline{y}_{\psi}^{*})}{\min M^{*}(T)}*100 = \frac{1+\sqrt{1-\rho_{yx}^2}}{(1-\rho_{yx}^2)\left[1+\sqrt{1-\rho_{yx.z}^2}\right]}*100$$

Remark. As the resulting MSE of the proposed generalized class $\min M^{*}(T)$ of chain type estimators coincides with the result of Singh and Vishwakarma [2], who provide the following inequality, which shows the relationship between the resulting mean square errors of Singh [5] & Singh and Vishwakarma [1]. Therefore, inequality will hold here, as well.

NUMERICAL STUDY

Since the above-discussed percentage relative efficiencies are independent of sample size, sampling variance, coefficient of variance, and other coefficients, all are based only on correlation coefficients. Therefore, we here provide numerical results in the following Table **1** for different admissible values of correlation coefficients and unmatched proportion.

Table 1. Percentage relative efficiency of a generalized class of chain type estimator $'T'$ at distinct values of ρ_{yx}, ρ_{xz} and ρ_{yz}.

ρ_{yx}		0.1		0.3		0.5		0.7		0.9	
ρ_{xz}	ρ_{yz}	E_1	E_2	E_1	E_2	E_1	E_2	E_1	E_2	E_1	E_2
	0.1	101.22	100.97	103.28	100.90	108.09	100.85	117.65	100.84	140.49	100.86
	0.3	110.04	109.76	112.21	109.62	117.61	109.73	128.65	110.76	157.01	112.72
	0.5	133.45	133.11	136.26	133.12	143.93	134.28	160.98	137.97	229.07	164.46
0.1	0.7	196.17	195.67	201.50	196.86	218..34	203.71	268.14	229.82	*	*
	0.9	526.38	525.07	561.42	548.49	793.79	740.62	*	*	*	*
	0.1	101.15	100.89	103.14	100.77	108.10	100.86	118.26	101.35	144.32	103.61
	0.3	109.89	109.62	111.39	108.83	116.12	108.34	126.16	108.13	150.97	108.39
	0.5	133.46	133.12	133.46	131.35	139.92	130.55	152.71	130.89	187.88	134.89
0.3	0.7	197.37	196.88	196.94	192.41	205.87	192.08	231.40	198.32	*	*
	0.9	550.36	548.99	527.00	514.87	574.24	535.77	*	*	*	*
	0.1	101.10	100.84	103.23	100.85	109.03	101.73	121.96	104.53	173.54	124.59
	0.3	103.99	109.72	110.81	108.26	115.32	107.60	125.86	107.87	154.85	111.17
	0.5	134.69	134.36	133.48	130.41	137.26	128.06	148.15	126.97	177.91	127.73
0.5	0.7	204.82	204.30	196.40	191.88	199.05	185.72	214.94	184.22	269.09	193.19
	0.9	765.75	763.83	548.91	536.27	528.64	493.23	601.76	515.75	*	*
	0.1	101.06	100.80	103.80	101.41	112.48	104.94	138.13	118.39	*	*
	0.3	110.62	110.34	110.37	107.83	115.38	107.65	129.68	111.15	*	*
	0.5	139.28	138.93	133.55	130.48	135.35	126.29	146.16	125.27	182.98	131.37
0.7	0.7	238.48	237.89	203.40	198.72	196.10	182.96	205.18	175.85	245.91	176.55
	0.9	*	*	*	*	551.51	514.57	533.14	456.94	702.84	504.60
	0.1	101.02	100.77	107.75	105.27	152.34	142.14	*	*	*	*
	0.3	114.91	114.62	110.03	107.50	119.90	111.86	*	*	*	*
	0.5	193.99	193.50	139.06	135.86	133.92	124.95	152.44	130.66	*	*
0.9	0.7	*	*	*	*	205.47	191.70	198.62	170.23	261.84	187.99
	0.9	*	*	*	*	*	*	579.85	496.97	559.70	401.84

Table 2. Optimum value of the unmatched proportion of generalized class of chain type estimator $'T'$ at different values of ρ_{yx}, ρ_{xz} and ρ_{yz}.

ρ_{yx}		0.1	0.3	0.5	0.7	0.9
ρ_{xz}	ρ_{yz}			θ_0		
0.1	0.1	0.5010	0.5112	0.5351	0.5824	0.6954
	0.3	0.5007	0.5105	0.5383	0.5854	0.7144
	0.5	0.5004	0.5110	0.5397	0.6037	0.8590
	0.7	0.5002	0.5138	0.5568	0.6838	*
	0.9	0.5000	0.5333	0.7541	*	*
0.3	0.1	0.5007	0.5105	0.5351	0.5854	0.7144
	0.3	0.5000	0.5068	0.5283	0.5740	0.6869
	0.5	0.5005	0.5042	0.5247	0.5727	0.7046
	0.7	0.5033	0.5022	0.5250	0.5901	*
	0.9	0.5228	0.5006	0.5455	*	*
0.5	0.1	0.5004	0.5110	0.5397	0.6037	0.8590
	0.3	0.5005	0.5042	0.5247	0.5727	0.7046
	0.5	0.5051	0.5006	0.5147	0.5556	0.6672
	0.7	0.5223	0.5008	0.5076	0.5481	0.6862
	0.9	0.7275	0.5215	0.5022	0.5717	*
0.7	0.1	0.5002	0.5138	0.5568	0.6838	*
	0.3	0.5033	0.5022	0.5250	0.5981	*
	0.5	0.5722	0.5008	0.5076	0.5481	0.6862
	0.7	0.6081	0.5187	0.5000	0.5232	0.6271
	0.9	*	*	0.5239	0.5065	0.6677
0.9	0.1	0.5000	0.5333	0.7541	*	*
	0.3	0.5228	0.5007	0.5455	*	*
	0.5	0.7275	0.5215	0.5022	0.5717	0.6964
	0.7	*	*	0.5239	0.5065	0.6677
	0.9	*	*	*	0.5509	0.5317

Table 3. Optimum unmatched proportion of estimator $\overline{y}_\psi$ or $\overline{y}_{\psi'}$.

ρ_{yx}	θ_1
0.1	0.5013
0.3	0.5118
0.5	0.5359
0.7	0.5834
0.9	0.6964

CONCLUSION AND INTERPRETATION

With an enormous potential in the sense of full coverage over several estimators, including the general classes and with the idea of the construction of classes, the proposed generalized class acquits an impressive position. The gain of the proposed study is computed in a dual sense, first is the case of no units matched from the earlier occasion and second without the use of auxiliary information at any occasion. We reckoned these gains by calculating efficiencies using the proportion of unmatched units obtained under the policy of optimum replacement and hence illustrated the efficiencies obtained in the above section.

From the obtained numerical results of the PRE with respect to the estimators $\bar{y}_n$, $\bar{y}_\psi$ and $\bar{y}_{\psi}$. in Table **1**, we observe that

i. For fixed ρ_{yx} and ρ_{xz} such that $\rho_{yx} = \rho_{xz}$, as we increase ρ_{yz}, efficiencies E_1 and E_2 are increasing.

ii. For fixed ρ_{yx} and ρ_{yz} such that $\rho_{yx} = \rho_{xz}$, as we increase ρ_{xz}, both the efficiencies are decreasing.

iii. For fixed ρ_{yz} and ρ_{xz} such that $\rho_{yz} = \rho_{xz}$, as we increase ρ_{yx}, calculated efficiencies are increasing.

iv. For $\rho_{yx} = \rho_{xz} = \rho_{yz} = \rho_0$, as we increase ρ_0 efficiencies are increasing as well.

CONSENT FOR PUBLICATION

Not applicable.

CONFLICT OF INTEREST

There is no conflict of interest declared.

ACKNOWLEDGEMENTS

The authors would like to express their heartfelt thanks to the honourable reviewer for his valuable suggestions.

REFERENCES

[1] H.P. Singh, and G.K. Vishwakarma, "A general class of estimators in successive sampling", *Metron,* vol. LXV, no. 2, pp. 201-227, 2007.

[2] H.P. Singh, and G.K. Vishwakarma, "A general procedure for estimating population mean in successive sampling", *Communications in Statistics- Theory and Methods,* vol. 38, pp. 293-308, 2009.

[http://dx.doi.org/10.1080/03610920802169594]

[3]   W.G. Cochran, *Sampling Techniques.*, 3rd ed Wiley Eastern Limited, 1977.

[4]   P.V. Sukhatme, B.V. Sukhatme, S. Sukhatme, and C. Asok, *Sampling Theory of Surveys with Applications.* Iowa State University Press: Ames, IA, 1984.

[5]   G.N. Singh, "On the use of chain-type ratio estimator in successive sampling", *Statist. Transition,* vol. 7, no. 1, pp. 21-26, 2005.

[6]   R.J. Jessen, "Statistical investigation of a survey for obtaining farm facts", *Iowa Agricult. Expt. Statist. Res. Bull,* vol. 304, pp. 1-104, 1942.

[7]   H. D. Patterson, "Sampling on successive occasions with partial replacement units", *J. Roy. Statist. Soc.,* vol. 12, no. B, pp. 241-255, 1950.
        [http://dx.doi.org/10.1111/j.2517-6161.1950.tb00058.x]

[8]   J. N. K. Rao, and J. E. Graham, "Rotation design for sampling on repeated occasions", *J. Amer. Statist. Assoc,* pp. 492-509.
        [http://dx.doi.org/10.1080/01621459.1964.10482175]

[9]   A.R. Sen, "Theory and applications of sampling on repeated occasions with several auxiliary variables", *Biometrics,* vol. 29, pp. 318-385, 1973.
        [http://dx.doi.org/10.2307/2529401]

[10]   P.C. Gupta, "Sampling on two successive occasions", *J. Stat. Res.,* vol. 13, pp. 7-13, 1979.

[11]   A.K. Das, "Estimation of population ratio on two occasions", *J. Ind. Soc. Agricult. Statist.,* vol. 34, pp. 1-9, 1982.

[12]   D.K. Chaturvedi, and T.P. Tripathi, "Estimation of population ratio on two occasions using multivariate auxiliary information", *J. Ind. Soc. Agricult. Statist.,* vol. 21, pp. 113-120, 1983.

[13]   S. Feng, and G. Zou, "Sample rotation method with auxiliary variable", *Commun. Stat. Theory Methods,* vol. 26, no. 6, pp. 1497-1509, 1997.
        [http://dx.doi.org/10.1080/03610929708831996]

[14]   R.S. Biradar, and H.P. Singh, "Successive sampling using auxiliary information on both the occasions", *Calcutta Statist. Assoc. Bull.,* vol. 51, pp. 243-251, 2001.
        [http://dx.doi.org/10.1177/0008068320010307]

[15]   G.N. Singh, and V.K. Singh, "On the use auxiliary information in successive sampling", *J. Ind. Soc. Agricult. Statist.,* vol. 54, no. 1, pp. 1-12, 2001.

[16]   G.N. Singh, "Estimation of population mean using auxiliary information on recent occasion in h^{th} occasions successive sampling", *Statist. Transition,* vol. 6, no. 4, pp. 523-532, 2003.

[17]   S. Bhushan, and N. Rastogi, "On a general class of estimators in double sampling using two auxiliary variables", *International Journal of Statistics and Systems,* vol. 13, no. 1, pp. 53-60, 2018.

[18]   L. Chand, Some ratio-type estimators based on two or more auxiliary variables Ph.D. dissertation, Iowa State University, Ames, Iowa, 1975.
        [http://dx.doi.org/10.31274/rtd-180814-485]

[19]   S.K. Srivastava, "A class of estimators using auxiliary information in sample surveys", *Canad. J. Statist.,* vol. 8, pp. 253-254, 1980.
        [http://dx.doi.org/10.2307/3315237]

[20]   J. Sahoo, and L.N. Sahoo, "A class of estimators in two phase sampling using two auxiliary variables", *Jour. Ind. Statist. Assoc.,* vol. 31, pp. 107-114, 1993.

[21]   B. Kiregyera, "A chain ratio-type estimator in finite population double sampling using two auxiliary variables", *Metrika,* vol. 27, no. 1, pp. 217-223, 1980.
        [http://dx.doi.org/10.1007/BF01893599]

[22]   B. Kiregyera, "Regression-type estimators using two auxiliary variables and the model of double sampling from finite populations", *Metrika,* vol. 31, no. 1, pp. 215-226, 1984.
        [http://dx.doi.org/10.1007/BF01915203]

CHAPTER 4

Log Type Estimators of Population Mean Under Ranked Set Sampling

Shashi Bhushan and **Anoop Kumar**[*]

Department of Mathematics and Statistics, Dr. Shakuntala Misra National Rehabilitation University, Lucknow, India

Abstract: This paper considers some log type and regression cum log type class of estimators under ranked set sampling. The suggested class of estimators are found to be better than most of the estimators proposed to date and equally efficient to the usual regression estimator under ranked set sampling. The theoretical findings have been furnished with a simulation study carried out over some artificially generated symmetric and asymmetric populations. Also, following McIntyre [1], Dell [2], and Dell and Clutter [3], we have investigated the effect of skewness and kurtosis over the efficiency of the proposed class of estimators.

Keywords: Bias, Efficiency, Kurtosis, Mean square error, Ranked set sampling, Skewness.

INTRODUCTION

McIntyre [1] mooted a cost-effective method in the theory of survey sampling and referred to it as a method of unbiased selective sampling using a ranked set, which consolidated the ease of non-probability sampling *via* curb of simple random sampling (SRS). It works as a stratification of samples rather than the classical approach, which preludes the stratification of the population. He showed theoretically that the mean of quantified units gives an unbiased estimator of a population mean irrespective of error in the ranking of units. However, the much expected stringent mathematical foundation to the theory of McIntyre's method was furnished by Takahasi and Wakimoto [4] and referred to it as a method of unbiased estimation of population mean on the sample stratified by means of ordering. The method of McIntyre [1] became inert nearly fourteen years until Halls and Dell [5] administered a field test for the estimation of weights of browse and herbage in a

[*]**Corresponding author Anoop Kumar:** Department of Mathematics and Statistics, Dr. Shakuntala Misra National Rehabilitation University, Lucknow, U.P., India; Tel: +91- 8009554263; E-mail: anoop.asy@gmail.com.

pine-hardwood forest. The first named it as ranked set sampling (RSS) and seen through the empirical observation that it was dominating the SRS. Dell and Clutter [3] pointed out that the RSS estimator of a population mean, is still unbiased with an equal sample size regardless of error involved during the judgement order. Muttlak and McDonald [6] developed RSS for the case when units are selected with size biased probability with respect to the concomitant variable. Muttlak and McDonald [7] evoked an efficient line intercept procedure under RSS. Muttlak [8] considered RSS for the estimation of parameters in simple linear regression. Samawi and Muttlak [9] showed, with the help of real data, which ranking of which variable increases the estimator's efficiency. Kadilar *et al.* [10] suggested a ratio estimator to estimate population mean under RSS. Al-Hadhrami [11] by following Kadilar and Cingi [12], suggested ratio type estimators of population mean under RSS. Al-Omari *et al.* [13] considered the new ratio estimators of population mean using SRS and RSS. Jeelani and Bouza [14] investigated a new ratio estimator using the linear combination of median and quartile deviation of the auxiliary variable under RSS, whereas Jeelani *et al.* [15] suggested ratio estimators of population mean utilizing the linear combination of deciles and coefficient of skewness of the auxiliary variable. Saini and Kumar [16] looked into the modified ratio estimator under SRS and RSS. Mehta and Mandowara [17] introduced modified ratio-cum-product estimators under RSS. Jeelani *et al.* [18] proposed a new ratio estimator under RSS based on deciles of an auxiliary variable. Khan *et al.* [19] considered a new regression cum ratio type estimator under RSS. Recently, Bhushan and Kumar [20] investigated some optimal classes of estimators under RSS. In this paper, we have adopted log type and regression cum log type class of estimators under RSS, employed by Bhushan *et al.* [21] and Bhushan and Gupta [22-26] under SRS. We have observed that in order to find out the expression of MSE, Al-Hadhrami [11] and Khan *et al.* [19] utilized the optimum value of optimizing scalar $\beta= \rho xy(Sy/Sx)$, an optimum value of b under SRS which is irrelevant for use in RSS. So, we have felt to derive the correct MSE by using the optimum value of b under RSS and comparing them with the proposed estimators class.

Further, section 2 is devoted to the methodology of ranked set sampling and review of some existing estimators under *RSS* in the literature. In section 3, the proposed estimators are given along with their properties. In section 4, we have derived the correct expression of *MSE* of Al-Hadhrami's [11] estimator and Khan *et al.*'s [19] estimator. In section 5, we have derived the theoretical conditions under which the proposed class of estimators dominates the other existing estimators. In section 6, a simulation study is added for an illustration. In section 7, the conclusion is made regarding the applicability of the adapted class of estimators. It is noticed that asymmetry shows an adverse effect on the efficiency of the proposed class of estimators. To verify this fact, we have computed the simulation results under the consideration of different asymmetric distributions.

LITERATURE REVIEW

The concept of ranked set sampling was given by McIntyre [1], which is based on picking out *m* simple random samples each of size *m* units from the population and m units are ranked within each set according to the variable of interest visually or by any cost independent measure. The unit now with *rank 1* is taken for the measurement of the element from the first sample, and the remaining units of the sample are discarded. Again, the unit with *rank 2* is taken for the measurement of the element from the second sample, and the remaining units of the sample are discarded. The process continues in the same way until the unit with *rank m* is taken for the measurement of the element from the mth sample. This above process is defined as a cycle. If this procedure recurred *r* times, then this yield a ranked set sample of size *n=mr* units. Let (X_1, Y_1), (X_2, Y_2),....,(X_n, Y_n) be a bivariate random sample of size *n* with probability density function *f (x, y)* and cumulative distribution function *F(x, y)* having population mean $\bar{X}$, $\bar{Y}$, variances σ_x^2, σ_y^2 and coefficient of correlation ρ_{xy}. Let the ranking be executed on the auxiliary variable x to estimate the population mean of the variable of interest y. Let (X_{11}, Y_{11}), (X_{12}, Y_{12}),...., (X_{1n}, Y_{1n}); (X_{21}, Y_{21}), (X_{22}, Y_{22}),...., (X_{2n}, Y_{2n});...;(X_{n1}, Y_{n1}), (X_{n2}, Y_{n2}),...., (X_{nn}, Y_{nn}) be a bivariate random sample drawn from the population with the help of simple random sampling without replacement *(SRSWOR)* having the same *c.d.f.* *F(x,y)* and $(X_{i(1)}, Y_{i[1]})$, $(X_{i(2)}, Y_{i[2]})$,...., $(X_{i(n)}, Y_{i[n]})$ be the order statistic of $X_{i1}, X_{i2},..., X_{in}$ and the judgment order of $Y_{i1}, Y_{i2},..., Y_{in}$; *i=1,2,...,n*. Let $X_{1(1)}, X_{2(2)},..., X_{n(n)}$ refers to the ranked set samples.

Samawi and Muttlak [9] suggested the estimator of population ratio as

$$\widehat{R}_r = \frac{\bar{y}_{[n]}}{\bar{x}_{(n)}} \tag{1}$$

Samawi and Muttlak [9] mentioned that the above estimator could also be utilized for the estimation of population mean as well as population total. Thus, the estimator for the population mean is

$$\bar{y}_r = \frac{\bar{y}_{[n]}}{\bar{x}_{(n)}} \bar{X} \tag{2}$$

Where $\bar{x}_{(n)} = \frac{1}{mr}\sum_{i=1}^{m} x_i$ and $\bar{y}_{[n]} = \frac{1}{mr}\sum_{i=1}^{m} y_i$ are the sample means of the auxiliary variable and study variable under *RSS*. The parenthesis () and [ ] used in the subscript of x and y, respectively, indicate the perfect ranking and imperfect ranking of units.

Yu and Lam [27] suggested a regression estimator under *RSS* as

$$t_\beta = \bar{y}_{[n]} + \beta\left(\bar{X} - \bar{x}_{(n)}\right) \tag{3}$$

Where β is the regression coefficient.

Motivated by Prasad [28] and Samawi and Muttlak [9], Kadilar *et al.* [10] suggested the following estimator as

$$\bar{y}_r = k\frac{\bar{y}_{[n]}}{\bar{x}_{(n)}}\bar{X} \tag{4}$$

where k is suitably considered constant.

Al-Omari *et al.* [13] suggested the ratio estimators of population mean as

$$\hat{\mu}_{y_h} = \bar{y}_{[n]}\left(\frac{\bar{X}+q_h}{\bar{x}_{(n)}+q_h}\right); h = 1,3 \tag{5}$$

where q_1 and q_3 represent the first and third quartiles of auxiliary variable x.

Jeelani and Bouza [14] suggested new ratio estimator comprises a linear combination of median and quartile deviation of the auxiliary variable as

$$\widehat{\bar{y}_\tau} = \tau\bar{y}_{[n]}\left(\frac{\bar{X}}{\bar{x}_{(n)}}\right) \tag{6}$$

Where $\tau = \left(\frac{\bar{X}Md+Qd}{\bar{x}_{(n)}Md+Q}\right)$, Md=median, Qd=quartile deviation=$\frac{(q_3-q_1)}{2}$

Saini and Kumar [16] suggested ratio type estimator as

$$\pi_h = \bar{y}_{[n]}\left(\frac{\bar{X}-\bar{x}_{(n)}+q_h}{\bar{X}+\bar{x}_{(n)}+q_h}\right); h = 1,3 \tag{7}$$

Jeelani *et al.* [15] suggested ratio estimators of population mean based on a linear combination of the population mean and decile of auxiliary variable as

$$\bar{y}_l = \bar{y}_{[n]} \left(\frac{\bar{X}+D_l}{\bar{x}_{(n)}+D_l} \right) \beta_1(x) \tag{8}$$

where D_l, l=1,2,...,9 are the deciles of the auxiliary variable.

Motivated by Sisodia and Dwivedi [29], Singh and Kakran [30], Upadhyaya and Singh [31] and Singh and Espejo [32], Mehta and Mandowara [17] investigated a class of ratio-cum-product type estimators as

$$\bar{y}_{MM1} = \bar{y}_{[n]} \left(\frac{\bar{X}+C_x}{\bar{x}_{(n)}+C_x} \right) \tag{9}$$

$$\bar{y}_{MM2} = \bar{y}_{[n]} \left(\frac{\bar{X}+\beta_2(x)}{\bar{x}_{(n)}+\beta_2(x)} \right) \tag{10}$$

$$\bar{y}_{MM3} = \bar{y}_{[n]} \left(\frac{\bar{X}C_x+\beta_2(x)}{\bar{x}_{(n)}C_x+\beta_2(x)} \right) \tag{11}$$

$$\bar{y}_{MM4} = \bar{y}_{[n]} \left(\frac{\bar{x}_{(n)}C_x+\beta_2(x)}{\bar{X}C_x+\beta_2(x)} \right) \tag{12}$$

$$\bar{y}_{MM5} = \bar{y}_{[n]} \left[\alpha \left(\frac{\bar{X}C_x+\beta_2(x)}{\bar{x}_{(n)}C_x+\beta_2(x)} \right) + (1-\alpha) \left(\frac{\bar{x}_{(n)}C_x+\beta_2(x)}{\bar{X}C_x+\beta_2(x)} \right) \right] \tag{13}$$

where α is a suitably chosen constant.

Jeelani *et al.* [18] suggested new ratio estimator consist of the decile of the auxiliary variable as

$$\hat{\bar{y}}_{\eta_1} = \eta_1 \bar{y}_{[n]} \left(\frac{\bar{X}}{\bar{x}_{(n)}} \right) \tag{14}$$

where $\eta_1 = \left(\frac{\bar{X}}{\bar{x}_{(n)}} + 1 \right) D_1$

The biases and mean square errors (*MSEs*) of these estimators are given in Appendix *I* for quick reference.

PROPOSED ESTIMATORS

Actuated by Bhushan *et al.* [21] and Bhushan and Gupta [22-26] we propose a new log type and regression cum log type class of estimators as

$$T_{G_1} = \bar{y}_{[n]} \left[1 + \log\left(\tfrac{\bar{x}_{(n)}}{\bar{X}}\right) \right]^{\alpha_1} \tag{15}$$

$$T_{G_2} = \bar{y}_{[n]} \left[1 + \alpha_2 \log\left(\tfrac{\bar{x}_{(n)}}{\bar{X}}\right) \right] \tag{16}$$

$$T_{G_3} = \left[\bar{y}_{[n]} + \alpha_3 (\bar{X} - \bar{x}_{(n)}) \right] \left[1 + \log\left(\tfrac{\bar{x}_{(n)}}{\bar{X}}\right) \right] \tag{17}$$

$$T_{G_4} = \bar{y}_{[n]} \left[1 + \log\left(\tfrac{\bar{x}_{(n)}{}^{*}}{\bar{X}^{*}}\right) \right]^{\alpha_4} \tag{18}$$

$$T_{G_5} = \bar{y}_{[n]} \left[1 + \alpha_5 \log\left(\tfrac{\bar{x}_{(n)}{}^{*}}{\bar{X}^{*}}\right) \right] \tag{19}$$

$$T_{G_6} = \left[\bar{y}_{[n]} + \alpha_6 (\bar{X} - \bar{x}_{(n)}) \right] \left[1 + \log\left(\tfrac{\bar{x}_{(n)}{}^{*}}{\bar{X}^{*}}\right) \right] \tag{20}$$

where $\alpha_j, j = 1,2,...,6$ are suitably chosen scalars, also $\bar{x}_{(n)}{}^{*} = a\bar{x}_{(n)} + b$ and $\bar{X}^{*} = a\bar{X} + b$ provided that $a(\neq 0)$, b is either real numbers or functions of known parameters of the auxiliary variables x like mean, coefficient of skewness $\beta_1(x)$, coefficient of kurtosis $\beta_2(x)$, coefficient of variation C_x, coefficient of correlation ρ_{xy}, decile D_1 and quartile q_h *etc.*

Theorem 1. *The biases of the proposed class of estimators are*

$$Bias(T_{G_1}) = \bar{Y} \left[\left(\tfrac{\alpha_1^2}{2} - \alpha_1\right)\left(\gamma C_x^2 - W_{x_{(i)}}^2\right) + \alpha_1 \left(\gamma \rho_{xy}\, C_x\, C_y - W_{xy_{[i]}}\right) \right] \tag{21}$$

$$Bias(T_{G_2}) = \bar{Y} \left[\alpha_2 \left(\gamma \rho_{xy}\, C_x\, C_y - W_{xy_{[i]}}\right) - \tfrac{\alpha_2}{2}\left(\gamma C_x^2 - W_{x_{(i)}}^2\right) \right] \tag{22}$$

$$Bias(T_{G_3}) = \bar{Y}$$

$$\left[\left(\gamma \rho_{xy}\, C_x\, C_y - W_{xy_{[i]}}\right) - \left(\tfrac{\alpha_3}{R} - \tfrac{1}{2}\right)\left(\gamma C_x^2 - W_{x_{(i)}}^2\right) \right] \tag{23}$$

$$Bias(T_{G_4}) = \bar{Y}\left[\left(\frac{\alpha_4^2}{2} - \alpha_4\right)\vartheta^2\left(\gamma C_x^2 - W_{x_{(i)}}^2\right) + \alpha_4\vartheta\left(\gamma\rho_{xy}\,C_x\,C_y - W_{xy_{[i]}}\right)\right] \quad (24)$$

$$Bias(T_{G_5}) = \bar{Y}\left[\alpha_5\vartheta\left(\gamma\rho_{xy}\,C_x\,C_y - W_{xy_{[i]}}\right) - \frac{\alpha_5}{2}\vartheta^2\left(\gamma C_x^2 - W_{x_{(i)}}^2\right)\right] \quad (25)$$

$$Bias(T_{G_6}) = \bar{Y}\left[\vartheta\left(\gamma\rho_{xy}\,C_x\,C_y - W_{xy_{[i]}}\right) - \left(\frac{\alpha_6}{R} - \frac{1}{2}\right)\vartheta^2\left(\gamma C_x^2 - W_{x_{(i)}}^2\right)\right] \quad (26)$$

Proof. To find out biases and *MSEs* of the proposed class of estimators, let us assume that

$$\bar{y}_{[n]} = \bar{Y}(1 + \varepsilon_0), \quad \bar{x}_{(n)} = \bar{X}(1 + \varepsilon_1), \quad \text{such that} \quad E(\varepsilon_0) = \quad E(\varepsilon_1) = 0 \quad \text{with}$$

$$E(\varepsilon_0^2) = \left(\gamma C_y^2 - W_{y_{[i]}}^2\right), E(\varepsilon_1^2) = \left(\gamma C_x^2 - W_{x_{(i)}}^2\right) \text{ and } E(\varepsilon_0\varepsilon_1) = \left(\gamma\rho_{xy}\,C_x\,C_y - W_{xy_{[i]}}\right)$$

Where $\tau_{x_{(i)}} = (\mu_{x_{(i)}} - \bar{X})$, $\tau_{y_{[i]}} = (\mu_{y_{[i]}} - \bar{Y})$, $\tau_{xy_{[i]}} = (\mu_{x_{(i)}} - \bar{X})(\mu_{y_{[i]}} - \bar{Y})$

Now, consider the estimator

$$T_{G_1} = \bar{y}_{[n]}\left[1 + \log\left(\frac{\bar{x}_{(n)}}{\bar{X}}\right)\right]^{\alpha_1} \quad (27)$$

Putting the value of $\bar{y}_{[n]}$ and $\bar{x}_{(n)}$ in the above equation and using Taylor series expansion, we get

$$T_{G_1} - \bar{Y} = \bar{Y}\left[\varepsilon_0 + \alpha_1\varepsilon_1 - \alpha_1\varepsilon_1^2 + \frac{\alpha_1^2}{2}\varepsilon_1^2 + \alpha_1\varepsilon_0\varepsilon_1\right] \quad (28)$$

Taking expectation both the sides of (28), we get the bias of the estimator T_{G_1} as

$$Bias(T_{G_1}) = \bar{Y}\left[\left(\frac{\alpha_1^2}{2} - \alpha_1\right)\left(\gamma C_x^2 - W_{x_{(i)}}^2\right) + \alpha_1\left(\gamma\rho_{xy}\,C_x\,C_y - W_{xy_{[i]}}\right)\right] \quad (29)$$

The bias of the remaining estimators of the *Theorem 1* can be obtained in similar lines.

Theorem 2. *The mean square errors of the proposed estimators are*

$$MSE\left(T_{G_j}\right) = \bar{Y}^2 \left[\begin{array}{c} \left(\gamma C_y^2 - W_{y_{[i]}}^2\right) + \alpha_j^2 \left(\gamma C_x^2 - W_{x_{(i)}}^2\right) \\ +2\alpha_j \left(\gamma \rho_{xy}\, C_x\, C_y - W_{xy_{[i]}}\right) \end{array} \right] ; j = 1, 2 \qquad (30)$$

$$MSE(T_{G_3}) = \bar{Y}^2 \left[\begin{array}{c} \left(\gamma C_y^2 - W_{y_{[i]}}^2\right) + \left(1 - \frac{\alpha_3}{R}\right)^2 \left(\gamma C_x^2 - W_{x_{(i)}}^2\right) \\ +2\left(1 - \frac{\alpha_3}{R}\right)\left(\gamma \rho_{xy}\, C_x\, C_y - W_{xy_{[i]}}\right) \end{array} \right] \qquad (31)$$

$$MSE\left(T_{G_j}\right) = \bar{Y}^2 \left[\begin{array}{c} \left(\gamma C_y^2 - W_{y_{[i]}}^2\right) + \alpha_j^2 \vartheta^2 \left(\gamma C_x^2 - W_{x_{(i)}}^2\right) \\ +2\alpha_j \vartheta \left(\gamma \rho_{xy}\, C_x\, C_y - W_{xy_{[i]}}\right) \end{array} \right] ; j = 4, 5 \qquad (32)$$

$$MSE(T_{G_6}) = \bar{Y}^2 \left[\begin{array}{c} \left(\gamma C_y^2 - W_{y_{[i]}}^2\right) + \left(\frac{\alpha_6}{R}\right)^2 \left(\gamma C_x^2 - W_{x_{(i)}}^2\right) + \vartheta^2 \left(\gamma C_x^2 - W_{x_{(i)}}^2\right) \\ -2\left(\frac{\alpha_6}{R}\right)\left(\gamma \rho_{xy}\, C_x\, C_y - W_{xy_{[i]}}\right) - 2\left(\frac{\alpha_6}{R}\right)\vartheta^2 \left(\gamma C_x^2 - W_{x_{(i)}}^2\right) \\ +2\vartheta \left(\gamma \rho_{xy}\, C_x\, C_y - W_{xy_{[i]}}\right) \end{array} \right] \qquad (33)$$

Proof. Squaring both the sides of (28) and taking expectation, we will get the *MSE* of the above estimator up to first-order approximation as

$$MSE(T_{G_1}) = \bar{Y}^2 \left[\begin{array}{c} \left(\gamma C_y^2 - W_{y_{[i]}}^2\right) + \alpha_1^2 \left(\gamma C_x^2 - W_{x_{(i)}}^2\right) \\ +2\alpha_1 \left(\gamma \rho_{xy}\, C_x\, C_y - W_{xy_{[i]}}\right) \end{array} \right] \qquad (34)$$

Similarly, the *MSEs* of the remaining estimators of *Theorem 2* can be ascertained.

Corollary 1. *The minimum MSE at the optimum value of* α_j *is*

$$minMSE\left(T_{G_j}\right) = \bar{Y}^2 \left[\left(\gamma C_y^2 - W_{y_{[i]}}^2\right) - \frac{\left(\gamma \rho_{xy}\, C_x\, C_y - W_{xy_{[i]}}\right)^2}{\left(\gamma C_x^2 - W_{x_{(i)}}^2\right)} \right] ; j = 1, 2, \dots, 6 \quad (35)$$

which is equal to the minimum *MSE* of classical regression estimator under *RSS* due to Yu and Lam [27], also it can be noticed that in the case of regression cum log type estimator, the minimum *MSE* does not depend upon the logarithmic part.

Correct Expressions of the *MSE* of Al-Hadhrami's [11] Estimator $\bar{y}_h$ and Khan *et al.*'s [19] Estimator $\bar{y}_p$

Following Kadilar and Cingi [12], Al-Hadhrami [11] suggested regression cum ratio estimators as

$$\bar{y}_h = [\bar{y}_{[n]} + \beta(\bar{X} - \bar{x}_{(n)})]\left(\frac{\alpha\bar{X}+\Gamma}{\alpha\bar{x}_{(n)}+\Gamma}\right) \tag{36}$$

where β is suitably chosen optimizing scalar, α and Γ are the parameters of auxiliary variable. Utilizing some known values of parameters of the auxiliary variables, he defined

$$\bar{y}_{h_1} = [\bar{y}_{[n]} + \beta(\bar{X} - \bar{x}_{(n)})]\left(\frac{\bar{X}}{\bar{x}_{(n)}}\right) \tag{37}$$

$$\bar{y}_{h_2} = [\bar{y}_{[n]} + \beta(\bar{X} - \bar{x}_{(n)})]\left(\frac{\bar{X}+C_x}{\bar{x}_{(n)}+C_x}\right) \tag{38}$$

$$\bar{y}_{h_3} = [\bar{y}_{[n]} + \beta(\bar{X} - \bar{x}_{(n)})]\left(\frac{\bar{X}+\beta_2(x)}{\bar{x}_{(n)}+\beta_2(x)}\right) \tag{39}$$

$$\bar{y}_{h_4} = [\bar{y}_{[n]} + \beta(\bar{X} - \bar{x}_{(n)})]\left(\frac{\bar{X}\beta_2(x)+C_x}{\bar{x}_{(n)}\beta_2(x)+C_x}\right) \tag{40}$$

$$\bar{y}_{h_5} = [\bar{y}_{[n]} + \beta(\bar{X} - \bar{x}_{(n)})]\left(\frac{\bar{X}C_x+\beta_2(x)}{\bar{x}_{(n)}C_x+\beta_2(x)}\right) \tag{41}$$

where C_x and $\beta_2(x)$ are the coefficient of variation and coefficient of kurtosis of auxiliary variable, respectively.

The minimum *MSE* of the estimator $\bar{y}_{h_u}$ deduced by Al-Hadhrami [11] is

$$MSE(\bar{y}_{h_u}) = \begin{bmatrix} \bar{Y}^2\left(\gamma C_y^2 - W_{y_{[i]}}^2\right) + D^2\bar{X}^2\left(\gamma C_x^2 - W_{x_{(i)}}^2\right) \\ -2D\bar{X}\bar{Y}\left(\gamma\rho_{xy}\,C_x\,C_y - W_{xy_{[i]}}\right) \end{bmatrix} \tag{42}$$

where $D = [\beta + \alpha\bar{Y}/(\alpha\bar{X} + \Gamma)]$ and $\beta = \rho_{xy}(S_y/S_x)$ is the optimum value of β in case of SRS which is not adequate to use in the case of RSS as it provides less efficient result as compared to the optimum value derived under RSS.

Now minimizing (42) w.r.t. β, we get

$$\beta^*_{(opt)} = \left[R \frac{\left(\gamma \rho_{xy}\, C_x\, C_y - W_{xy_{[i]}}\right)}{\left(\gamma C_x^2 - W_{x_{(i)}}^2\right)} - \left(\frac{\alpha \bar{Y}}{\alpha \bar{X} + \Gamma}\right) \right] \tag{43}$$

Putting the optimum value of β in (42), we get minimum *MSE* as

$$minMSE(\bar{y}^*_{h_u}) = \bar{Y}^2 \left[\left(\gamma C_y^2 - W_{y_{[i]}}^2\right) - \frac{\left(\gamma \rho_{xy}\, C_x\, C_y - W_{xy_{[i]}}\right)^2}{\left(\gamma C_x^2 - W_{x_{(i)}}^2\right)} \right]; \; u = 1, 2, \dots, 5 \tag{44}$$

which is minimum *MSE* in case of classical regression estimator under *RSS* and independent of the value of $\left(\frac{\alpha \bar{X} + \Gamma}{\alpha \bar{x}_{(n)} + \Gamma}\right)$. Thus, the value of $\left(\frac{\alpha \bar{X} + \Gamma}{\alpha \bar{x}_{(n)} + \Gamma}\right)$ does not affect the $minMSE\left(\bar{y}_{h_u}\right)$ but affect the optimum value of β^* as it changes.

Khan *et al.* [19] proposed a new class of regression cum ratio estimator as

$$\bar{y}_p = [\bar{y}_{[n]} + b\left(\bar{X} - \bar{x}_{(n)}\right)] \left(\frac{\bar{A}\lambda + \delta}{\bar{a}_k + \delta}\right) \tag{45}$$

where λ and δ are known constants such as coefficient of variation, coefficient of skewness $\beta_1(x)$, coefficient of kurtosis $\beta_2(x)$, coefficient of correlation ρ_{xy}, lower quartile $q_1(x)$, upper quartile $q_3(x)$ *etc.*

Also b $= \frac{\bar{Y} \rho_{xy}\, C_y}{\bar{X} C_x}$ is the sample estimate of population regression coefficient and $a_k = \bar{x}_{(n)} + \bar{X}(b_k - 1)$, $\bar{A} = \bar{X} b_{k,}$ $k{=}1,2,\dots,6$ such that $b_1 = C_x$, $b_2 = \beta_1(x)$, $b_3 = \beta_2(x)$, $b_4 = \rho_{xy}$, $b_5 = q_1(x)$, $b_6 = q_3(x)$.

For different values of λ and δ, they defined a class of estimators as

$$\bar{y}_{p1} = [\bar{y}_{[n]} + b\left(\bar{x}_{(n)} - \bar{X}\right)] \left(\frac{\bar{A} + C_x}{\bar{a}_k + C_x}\right) \tag{46}$$

$$\bar{y}_{p2} = [\bar{y}_{[n]} + b\left(\bar{x}_{(n)} - \bar{X}\right)] \left(\frac{\bar{A} + \beta_2(x)}{\bar{a}_k + \beta_2(x)}\right) \tag{47}$$

$$\bar{y}_{p3} = [\bar{y}_{[n]} + b\left(\bar{x}_{(n)} - \bar{X}\right)] \left(\frac{\bar{A}\beta_2(x) + C_x}{\bar{a}_k \beta_2(x) + C_x}\right) \tag{48}$$

$$\bar{y}_{p_4} = \left[\bar{y}_{[n]} + b\left(\bar{x}_{(n)} - \bar{X}\right)\right] \left(\frac{\bar{A}C_x + \beta_2(x)}{\bar{a}_k C_x + \beta_2(x)}\right) \tag{49}$$

The bias and *MSE* of the estimator $\bar{y}_{p_j}, j = 1,2,3,4$ derived by Khan *et al.* [19] is

$$Bias\,(\bar{y}_{p_j}) = \bar{Y}\left[\left(\varphi_j^2 + \varphi_j \rho_{xy}\frac{C_y}{C_x}\right)\left(\gamma C_x^2 - W_{x_{(i)}}^2\right) - \varphi_j\left(\gamma\rho_{xy}\,C_x\,C_y - W_{xy_{[i]}}\right)\right] \tag{50}$$

$$MSE\left(\bar{y}_{p_j}\right) = \bar{Y}^2\left[\begin{array}{c}\left(\gamma C_y^2 - W_{y_{[i]}}^2\right) + \left(\varphi_{j+}\rho_{xy}\frac{C_y}{C_x}\right)^2 \\ -\left(\varphi_{j+}\rho_{xy}\frac{C_y}{C_x}\right)\left(\gamma\rho_{xy}\,C_x\,C_y - W_{xy_{[i]}}\right)\end{array}\right] \tag{51}$$

Where $\varphi_1 = (\bar{X}|(\bar{X}b_k + C_x))$, $\varphi_2 = (\bar{X}|(\bar{X}b_k + \beta_2(x)))$,
$\varphi_3 = (\bar{X}\beta_2(x)|(\bar{X}b_k\beta_2(x) + C_x))$, $\varphi_4 = (\bar{X}C_x|(\bar{X}b_k C_x + \beta_2(x)))$

Khan *et al.* [19] used $b = \dfrac{\bar{Y}\rho_{xy}\,C_y}{\bar{X}C_x}$ as the optimum value which is not appropriate and if use provides less efficient result as compare to the optimum value obtained under *RSS*.

Now consider the estimator in terms of ε_0 and ε_1, we get

$$\bar{y}_{p_j}^* - \bar{Y} = \bar{Y}\left[\varepsilon_0 - \frac{b}{R}\varepsilon_1 - \varphi_j\varepsilon_1 - \varphi_j\varepsilon_0\varepsilon_1 + \frac{b}{R}\varphi_j\varepsilon_1^2 + \varphi_j^2\varepsilon_1^2\right] \tag{52}$$

Taking expectation both the sides of (52), we get the bias of the estimator $\bar{y}_{p_j}^*$ as

$$Bias(\bar{y}_{p_j}^*) = \bar{Y}\left[\begin{array}{c}\frac{b}{R_x}\varphi_j\left(\gamma C_x^2 - W_{x_{(i)}}^2\right) + \varphi_j^2\left(\gamma C_x^2 - W_{x_{(i)}}^2\right) \\ -\varphi_j\left(\gamma\rho_{xy}\,C_x\,C_y - W_{xy_{[i]}}\right)\end{array}\right] \tag{53}$$

Now squaring and taking expectation both the sides of (52), we get

$$MSE\left(\bar{y}_{p_j}^*\right) = \bar{Y}^2\left[\begin{array}{c}\left(\gamma C_y^2 - W_{y_{[i]}}^2\right) + \frac{b}{R^2}^2\left(\gamma C_x^2 - W_{x_{(i)}}^2\right) + \varphi_j^2\left(\gamma C_x^2 - W_{x_{(i)}}^2\right) \\ -2\frac{b}{R}\left(\gamma\rho_{xy}\,C_x\,C_y - W_{xy_{[i]}}\right) + 2\frac{b}{R}\varphi_j\left(\gamma C_x^2 - W_{x_{(i)}}^2\right) \\ -2\varphi_j\left(\gamma\rho_{xy}\,C_x\,C_y - W_{xy_{[i]}}\right)\end{array}\right] \tag{54}$$

Now minimizing (54) w.r.t. b, we get

$$b_{(opt)} = R \left[\frac{\left(\gamma \rho_{xy} \, C_x \, C_y - W_{xy_{[i]}} \right)}{\left(\gamma C_x^2 - W_{x_{(i)}}^2 \right)} - \varphi_j \right] = b_{(opt)}^* (say) \tag{55}$$

Putting the optimum value of b in (54), we get

$$minMSE \left(\bar{y}_{p_j}^* \right) = \bar{Y}^2 \left[\left(\gamma C_y^2 - W_{y_{[i]}}^2 \right) - \frac{\left(\gamma \rho_{xy} \, C_x \, C_y - W_{xy_{[i]}} \right)^2}{\left(\gamma C_x^2 - W_{x_{(i)}}^2 \right)} \right]; \, j = 1, 2, 3, 4 \tag{56}$$

which is minimum *MSE* of classical regression estimator under *RSS* and it does not depend on the value of φ_j. Thus, whatever be the value of φ_j, the $minMSE \left(\bar{y}_{p_j}^* \right)$ is unaltered but the optimum value of b^* changes as the value of φ_j changes.

Theoretical Comparison

In this section, we have compared the proposed estimators with the estimators considered in this study. We get the following conditions.

1. From (35) and (72)

$$MSE(\bar{y}_m) > MSE \left(T_{G_j} \right)$$

$$\frac{\left(\gamma \rho_{xy} \, C_x \, C_y - W_{xy_{[i]}} \right)^2}{\left(\gamma C_x^2 - W_{x_{(i)}}^2 \right)} > -1 \tag{57}$$

2. From (35) and (73)

$$MSE(\bar{y}_r) > MSE \left(T_{G_j} \right)$$

$$\frac{\left(\gamma \rho_{xy} \, C_x \, C_y - W_{xy_{[i]}} \right)^2}{\left(\gamma C_x^2 - W_{x_{(i)}}^2 \right)} > \left(2 \left(\gamma \rho_{xy} \, C_x \, C_y - W_{xy_{[i]}} \right) - \left(\gamma C_x^2 - W_{x_{(i)}}^2 \right) \right) \tag{58}$$

3. From (35) and (77)

$$MSE(\bar{y}_k) > MSE \left(T_{G_j} \right)$$

$$\frac{\left(\gamma\rho_{xy}\,C_x\,C_y - W_{xy_{[i]}}\right)^2}{\left(\gamma C_x^2 - W_{x_{(i)}}^2\right)} > \left(\begin{array}{c} 2k^*\left(\gamma\rho_{xy}\,C_x\,C_y - W_{xy_{[i]}}\right) - \left(\gamma C_x^2 - W_{x_{(i)}}^2\right) \\ -\left(k^{*2} - 1\right)\left(\gamma C_y^2 - W_{y_{[i]}}^2\right) \end{array}\right) \tag{59}$$

4. From (35) and (78)

$$MSE(\hat{\mu}_{y_h}) > MSE\left(T_{G_j}\right)$$

$$\frac{\left(\gamma\rho_{xy}\,C_x\,C_y - W_{xy_{[i]}}\right)^2}{\left(\gamma C_x^2 - W_{x_{(i)}}^2\right)} > \left(\begin{array}{c} \left(\gamma C_y^2 - W_{y_{[i]}}^2\right) - \gamma C_y^2(1 - \rho_{xy}^2) \\ -\frac{(K_h - \beta)^2}{R^2}\left(\gamma C_x^2 - W_{x_{(i)}}^2\right) \end{array}\right) \tag{60}$$

5. From (35) and (79)

$$MSE(\hat{\bar{y}}_\tau) > MSE\left(T_{G_j}\right)$$

$$\frac{\left(\gamma\rho_{xy}\,C_x\,C_y - W_{xy_{[i]}}\right)^2}{\left(\gamma C_x^2 - W_{x_{(i)}}^2\right)} > \left(\begin{array}{c} 2\tau\left(\gamma\rho_{xy}\,C_x\,C_y - W_{xy_{[i]}}\right) - (\tau^2 - 1)\left(\gamma C_y^2 - W_{y_{[i]}}^2\right) \\ -\left(\gamma C_x^2 - W_{x_{(i)}}^2\right) - (1 - 2\tau)^2 \end{array}\right) \tag{61}$$

6. From (35) and (80)

$$MSE(\pi_h) > MSE\left(T_{G_j}\right)$$

$$\frac{\left(\gamma\rho_{xy}\,C_x\,C_y - W_{xy_{[i]}}\right)^2}{\left(\gamma C_x^2 - W_{x_{(i)}}^2\right)} > \left(2\frac{I_h}{R}\left(\gamma\rho_{xy}\,C_x\,C_y - W_{xy_{[i]}}\right) - \left(\frac{I_h}{R}\right)^2\left(\gamma C_x^2 - W_{x_{(i)}}^2\right)\right) \tag{62}$$

7. From (35) and (81)

$$MSE(\bar{y}_l) > MSE\left(T_{G_j}\right)$$

$$\frac{\left(\gamma\rho_{xy}\,C_x\,C_y - W_{xy_{[i]}}\right)^2}{\left(\gamma C_x^2 - W_{x_{(i)}}^2\right)} > \left(\begin{array}{c} \left(\gamma C_y^2 - W_{y_{[i]}}^2\right) - \gamma C_y^2(1 - \rho_{xy}^2) \\ -\frac{(Z_l - \beta)^2}{R^2}\left(\gamma C_x^2 - W_{x_{(i)}}^2\right) \end{array}\right) \tag{63}$$

8. From (35) and (82)

$$MSE(\bar{y}_{MM_i}) > MSE\left(T_{G_j}\right), i = 1,2,3$$

$$\frac{\left(\gamma\rho_{xy}\, C_x\, C_y - W_{xy_{[i]}}\right)^2}{\left(\gamma C_x^2 - W_{x_{(i)}}^2\right)} > \left(2\gamma_i\left(\gamma\rho_{xy}\, C_x\, C_y - W_{xy_{[i]}}\right) - \gamma_i^2\left(\gamma C_x^2 - W_{x_{(i)}}^2\right)\right) \tag{64}$$

9. From (35) and (83)

$$MSE(\bar{y}_{MM_4}) > MSE\left(T_{G_j}\right)$$

$$\frac{\left(\gamma\rho_{xy}\, C_x\, C_y - W_{xy_{[i]}}\right)^2}{\left(\gamma C_x^2 - W_{x_{(i)}}^2\right)} > \left(\gamma_i^2\left(\gamma C_x^2 - W_{x_{(i)}}^2\right) - 2\gamma_i\left(\gamma\rho_{xy}\, C_x\, C_y - W_{xy_{[i]}}\right)\right) \tag{65}$$

10. From (35) and (85)

$$MSE(\bar{y}_{MM_5}) > MSE\left(T_{G_j}\right)$$

$$\frac{\left(\gamma\rho_{xy}\, C_x\, C_y - W_{xy_{[i]}}\right)^2}{\left(\gamma C_x^2 - W_{x_{(i)}}^2\right)} > \left(\begin{array}{c}\left(\gamma C_y^2 - W_{y_{[i]}}^2\right) - \gamma S_y^2(1 - \rho_{xy}^2) \\ + \left(W_{y_{[i]}} - \left(\frac{t_3 - K}{2}\right)W_{x_{(i)}}\right)^2\end{array}\right) \tag{66}$$

11. From (35) and (86)

$$MSE(\hat{\bar{y}}_{\eta l}) > MSE\left(T_{G_j}\right)$$

$$\frac{\left(\gamma\rho_{xy}\, C_x\, C_y - W_{xy_{[i]}}\right)^2}{\left(\gamma C_x^2 - W_{x_{(i)}}^2\right)} > \left(2\eta_l\left(\gamma\rho_{xy}\, C_x\, C_y - W_{xy_{[i]}}\right) - \eta_l^2\left(\gamma C_x^2 - W_{x_{(i)}}^2\right)\right) \tag{67}$$

12. From (35) and (42)

$$MSE(\bar{y}_h) > MSE\left(T_{G_j}\right)$$

$$\frac{\left(\gamma\rho_{xy}\, C_x\, C_y - W_{xy_{[i]}}\right)^2}{\left(\gamma C_x^2 - W_{x_{(i)}}^2\right)} > \frac{D}{R}\left(2\left(\gamma\rho_{xy}\, C_x\, C_y - W_{xy_{[i]}}\right) - \frac{D}{R}\left(\gamma C_x^2 - W_{x_{(i)}}^2\right)\right) \tag{68}$$

13. From (35) and (51)

$$MSE\left(\bar{y}_{P_j}\right) > MSE\left(T_{G_j}\right)$$

$$\frac{\left(\gamma\rho_{xy}\, C_x\, C_y - W_{xy_{[i]}}\right)^2}{\left(\gamma C_x^2 - W_{x_{(i)}}^2\right)} > \left(\begin{array}{c} 2\left(\varphi_{j} + \rho_{xy}\dfrac{C_y}{C_x}\right)\left(\gamma\rho_{xy}\, C_x\, C_y - W_{xy_{[i]}}\right) \\[2mm] -\left(\varphi_{j} + \rho_{xy}\dfrac{C_y}{C_x}\right)^2\left(\gamma C_x^2 - W_{x_{(i)}}^2\right) \end{array}\right) \tag{69}$$

Simulation Study

In order to gain some insight concerning to the efficiency as well as the impact of skewness and kurtosis over the efficiency of the proposed class of estimators, following Singh and Horn [33], a simulation study is performed over hypothetically generated symmetric and asymmetric populations having population size $N = 1000$ units with variables X and Y obtained through the following transformation given as

$$y_i = 2.8 + \sqrt{\left(1 - \rho_{xy}^2\right)}\, y_i^* + \rho_{xy}\left(\frac{S_y}{S_x}\right) x_i^* \tag{70}$$

and

$$x_i = 2.4 + x_i^* \tag{71}$$

where x_i^* and y_i^* are independent variates of corresponding parent distributions. The sampling methodology discussed in the earlier section is used to obtain a ranked set sample of size 12 units with set size 3 and a number of cycles 4 from each population. Using 10000 replications, the percent relative efficiency (*PRE*) of the proposed estimators with respect to (w.r.t.) the conventional mean estimator is computed as

$$PRE = \frac{MSE\left(\bar{y}_m\right)}{MSE\left(T\right)} \times 100$$

The results of the simulation experiments, which shows the dominance of proposed estimators with other existing estimators are summarized hereunder in Tables **1-3** by the *PRE* for each sensibly preferred values of the correlation coefficient $\rho_{xy} = 0.6, 0.7, 0.8, 0.9$. Also, the results which show the effect of skewness over the *PRE* of the suggested class of estimators $T = T_{G_j}, t_\beta, \bar{y}^*_{h_u}, \bar{y}^*_{p_j}$ w.r.t. conventional mean estimator $\bar{y}_m$ are reported in Tables **4-7** for $\rho_{xy} = 0.9$.

Results of the Simulation Study

On the basis of Tables **1-7** we may infer that

1. The *PRE* of the proposed class of estimators $T_{G_j}, j = 1, 2, \ldots, 6$ after correction, Al-Hadhrami's [11] estimator $\bar{y}^*_{h_u}, u = 1, 2, \ldots, 5$ and Khan *et al.'s* [19] estimator $\bar{y}^*_{p_j}, j = 1, 2, 3, 4$ are equally efficient to the classical regression estimator under *RSS* and more efficient to the other reviewed estimators for the chosen values of the correlation coefficient ρ_{xy}.

2. It can also be seen that the *PRE* of the corrected Al-Hadhrami's [11] estimator $\bar{y}^*_{h_u}, u = 1, 2, \ldots, 5$ and Khan *et al.'s* [19] estimator $\bar{y}^*_{p_j}, j = 1, 2, 3, 4$ are more efficient than their estimators $\bar{y}_{h_u}, u = 1, 2, \ldots, 5$ and $\bar{y}_{p_j}, j = 1, 2, 3, 4$ before correction.

3. The *PRE* of the suggested class of estimators is decreased as the correlation coefficient ρ_{xy} increases in case of *Normal* and *Weibull* population while increased as the correlation coefficient ρ_{xy} increases in the case of the *Chi-square* population, which is due to the adverse effect of skewness and kurtosis reported in the Tables **1-3** for corresponding populations.

4. Subsequently, we have also studied the effect of skewness and kurtosis over *PRE* of the suggested class of estimators $T = T_{G_j}, t_\beta, \bar{y}^*_{h_u}, \bar{y}^*_{p_j}$ w.r.t. conventional mean estimator $\bar{y}_m$ under different skewed populations for a reasonable choice of $\rho_{xy} = 0.9$ and noticed that *PRE* decreased as skewness and kurtosis increases. This tendency is sustained here by the simulation results computed for *Gamma, Weibull, Beta-I,* and *Chi-square* population reported in Tables **4-7**.

5. Our findings are also inconsonant with McIntyre [1], who observed that the efficiency of estimation of mean decreases with the tendency of distribution to depart from symmetry. The simulation findings were also obtained by Dell [2] and Dell and Clutter [3], who studied this issue over a wide range of skewed populations.

CONCLUSION

We have proposed some log type and regression cum log type class of estimators along with their properties. It has been seen that the minimum *MSE* of the proposed class of estimators along with the corrected Al-Hadhrami's [11] estimator and Khan *et al.'s* [19] estimator are equivalent to the minimum *MSE* of the classical regression estimator under *RSS*. The theoretical claims are enhanced by the results of the simulation study conducted on hypothetically generated symmetric and asymmetric populations. Likewise McIntyre [1], Dell [2] and Dell and Clutter [3], we have also studied the effect of skewness and kurtosis over *PRE* of the proposed class of estimators $T = T_{G_j},\ t_\beta,\ \bar{y}^*_{h_u},\ \bar{y}^*_{p_j}$ w.r.t. conventional mean estimator $\bar{y}_m$ under different skewed (*viz. Gamma, Weibull, Beta-I,* and *Chi-square*) populations. We have reported that *PRE* has an inverse numerical relationship with skewness and kurtosis. So it is worth pointing out that skewness and kurtosis should be kept in mind while dealing with an asymmetric population under *RSS*. It is seen theoretically and empirically that the suggested class of estimators is equally efficient to the classical regression estimator and better than the other existing estimators.

Table 1. *PRE* **of proposed estimators w.r.t. conventional mean estimator for** *Normal* **population.**

ρ_{xy} / Estimators	0.6	0.7	0.8	0.9
$x^* \sim N(12,5)$				
$y^* \sim N(15,7)$	0.0046	0.0110	0.0201	0.0338
Skewness(y)	2.9251	2.9604	2.9971	3.0194
Kurtosis(y)	**101.968**	**101.287**	**100.643**	**100.114**
$T_{G_j},\ j = 1,2,\dots,6$	100.000	100.000	100.000	100.000
$\bar{y}_m$	53.917	51.256	48.810	46.683
$\bar{y}_r$	101.968	101.287	100.643	100.114
t_β	54.063	51.391	48.940	46.816
$\bar{y}_k$	50.155	48.987	48.205	48.196
$\bar{y}_{h_1}$	59.630	58.412	57.570	57.491
$\bar{y}_{h_2}$	49.328	48.168	47.392	47.388
	74.344	73.163	72.290	72.067
$\bar{y}_{h_3}$	48.973	47.816	47.043	47.041
$\bar{y}_{h_4}$	101.986	101.287	100.643	100.114
$\bar{y}_{h_5}$	82.891	79.967	76.950	73.790
$\bar{y}^*_{h_u},\ u = 1,2,\dots,5$	90.420	87.793	84.965	81.827
	50.436	47.974	45.783	44.007
$\hat{\mu}_{y_1}$	90.175	87.534	84.695	81.551
$\hat{\mu}_{y_3}$	92.101	89.579	86.835	83.742
$\hat{\bar{y}}_\tau$	87.801	85.042	82.119	78.944
π_1	97.573	95.569	93.283	90.548
π_3	98.875	97.059	94.958	92.394

(Table 1) cont.....

$\bar{y}_1$	99.512	97.808	95.818	93.363
$\bar{y}_2$	99.937	98.316	96.410	94.041
$\bar{y}_3$	100.303	98.762	96.939	94.655
$\bar{y}_4$	100.654	99.199	97.466	95.277
$\bar{y}_5$	101.007	99.651	98.025	95.950
$\bar{y}_6$	101.785	100.763	99.505	97.847
$\bar{y}_7$	48.252	45.735	43.463	41.545
$\bar{y}_8$	58.190	55.335	52.667	50.264
$\bar{y}_9$	73.867	70.787	67.731	64.705
$\bar{y}_{MM_1}$	73.867	70.787	67.731	64.705
$\bar{y}_{MM_2}$	101.770	101.092	100.455	99.942
$\bar{y}_{MM_3}$	97.911	95.951	93.709	91.014
$\bar{y}_{MM_4}$	1.458	1.395	1.353	1.348
$\bar{y}_{MM_5}$	0.853	0.817	0.794	0.793
$\hat{\bar{y}}_{\eta_1}$	0.638	0.612	0.595	0.595
$\hat{\bar{y}}_{\eta_2}$	0.517	0.495	0.482	0.482
$\hat{\bar{y}}_{\eta_3}$	0.423	0.406	0.395	0.396
$\hat{\bar{y}}_{\eta_4}$	0.342	0.329	0.320	0.321
$\hat{\bar{y}}_{\eta_5}$	0.268	0.257	0.250	0.251
$\hat{\bar{y}}_{\eta_6}$	0.114	0.109	0.107	0.107
$\hat{\bar{y}}_{\eta_7}$	8.831	8.543	8.380	8.462
$\hat{\bar{y}}_{\eta_8}$	19.141	18.591	18.285	18.460
$\hat{\bar{y}}_{\eta_9}$	8.131	7.863	7.712	7.787
$\bar{y}_{p_1}$	43.385	42.485	41.983	42.290
$\bar{y}_{p_2}$	101.968	101.287	100.643	100.114
$\bar{y}_{p_3}$				
$\bar{y}_{p_4}$				
$\bar{y}^*_{p_j}, j=1,2,3,4$				

where * represents the corrected *PRE* values of estimators due to Al-Hadhrami [11] and Khan *et al.* [19].

Table 2. *PRE* of proposed estimators w.r.t. conventional mean estimator for *Weibull* population.

ρ_{xy} / Estimators	0.6	0.7	0.8	0.9
$x^* \sim Wb(2,1)$				
$y^* \sim Wb(2,2)$	0.3376	0.3274	0.3618	0.4592
Skewness(y)	2.9652	2.9903	3.0567	3.1746
Kurtosis(y)	**191.232**	**188.312**	**179.421**	**158.393**
$T_{G_j}, j=1,2,\dots,6$	100.000	100.000	100.000	100.000
$\bar{y}_m$	186.138	176.654	156.166	118.805
$\bar{y}_r$	191.232	188.312	179.421	158.393
t_β	186.280	176.746	156.216	118.827
$\bar{y}_k$	100.966	94.853	86.184	73.874
$\bar{y}_{h_1}$	154.354	150.059	141.278	124.384
$\bar{y}_{h_2}$	97.801	91.679	83.114	71.118
	190.653	187.984	179.306	158.390

(Table 2) cont.....

Estimator				
$\bar{y}_{h_3}$	95.928	89.806	81.308	69.501
$\bar{y}_{h_4}$	191.232	188.312	179.421	158.393
$\bar{y}_{h_5}$	168.925	172.013	171.368	157.901
$\bar{y}_{h_u}^*,\, u = 1, 2, \ldots, 5$	162.810	166.073	166.480	156.110
	186.113	176.633	156.151	118.797
$\hat{\mu}_{y_1}$	159.231	162.507	163.361	154.562
$\hat{\mu}_{y_3}$	157.330	160.591	161.640	153.610
$\hat{\bar{y}}_\tau$	154.351	157.563	158.864	151.949
π_1	151.101	154.229	155.738	149.923
π_3	149.615	152.694	154.277	148.926
$\bar{y}_1$	148.403	151.439	153.073	148.083
$\bar{y}_2$	147.147	150.135	151.814	147.181
$\bar{y}_3$	146.016	148.957	150.670	146.346
$\bar{y}_4$	144.736	147.621	149.366	145.374
$\bar{y}_5$	142.877	145.675	147.450	143.915
$\bar{y}_6$	135.644	138.052	139.819	137.752
$\bar{y}_7$	169.067	150.438	121.393	82.535
$\bar{y}_8$	171.055	171.852	163.423	133.838
$\bar{y}_9$	113.888	117.562	121.072	121.214
$\bar{y}_{MM_1}$	113.888	117.562	121.072	121.214
$\bar{y}_{MM_2}$	188.343	184.204	173.875	151.674
$\bar{y}_{MM_3}$	84.114	71.007	56.058	39.666
$\bar{y}_{MM_4}$	63.111	52.813	41.565	29.626
$\bar{y}_{MM_5}$	55.109	46.014	36.217	25.923
$\hat{\bar{y}}_{\eta_1}$	49.280	41.098	32.365	23.251
$\hat{\bar{y}}_{\eta_2}$	43.845	36.537	28.800	20.772
$\hat{\bar{y}}_{\eta_3}$	39.436	32.853	25.924	18.765
$\hat{\bar{y}}_{\eta_4}$	34.955	29.119	23.012	16.726
$\hat{\bar{y}}_{\eta_5}$	29.300	24.419	19.345	14.145
$\hat{\bar{y}}_{\eta_6}$	14.523	12.165	9.745	7.284
$\hat{\bar{y}}_{\eta_7}$	3.923	3.452	2.951	2.426
$\hat{\bar{y}}_{\eta_8}$	47.072	44.929	42.148	38.600
$\hat{\bar{y}}_{\eta_9}$	2.901	2.545	2.168	1.775
$\bar{y}_{p_1}$	113.090	116.638	119.833	120.006
$\bar{y}_{p_2}$	191.232	188.312	179.421	158.393
$\bar{y}_{p_3}$				
$\bar{y}_{p_4}$				
$\bar{y}_{p_j}^*,\, j = 1, 2, 3, 4$				

Where * represents the corrected *PRE* values of estimators due to Al-Hadhrami [11] and Khan *et al.* [19].

Table 3. *PRE* of proposed estimators w.r.t. conventional mean estimator for χ^2 population.

ρ_{xy} Estimators	0.6	0.7	0.8	0.9
$x^* \sim \chi^2(7)$				
$y^* \sim \chi^2(2)$	1.6356	1.3620	1.1092	0.9229
Skewness(y)	8.1349	6.5737	5.1771	4.2160

(Table 3) cont.....

Kurtosis(y)	**139.404**	**151.198**	**169.732**	**203.713**
$T_{G_j},\ j=1,2,...,6$	100.000	100.000	100.000	100.000
$\bar{y}_m$	124.582	132.299	143.077	158.910
$\bar{y}_r$	139.404	151.198	169.732	203.713
t_β	125.288	132.803	143.378	159.016
$\bar{y}_k$	81.391	79.977	77.804	74.146
$\bar{y}_{h_1}$	90.819	90.692	90.109	88.398
$\bar{y}_{h_2}$	80.117	78.556	76.210	72.358
$\bar{y}_{h_3}$	105.280	107.791	110.839	114.358
$\bar{y}_{h_4}$	78.224	76.454	73.867	69.755
$\bar{y}_{h_5}$	139.404	151.198	169.732	203.713
	138.847	150.211	168.049	200.851
$\bar{y}^{*}_{h_u},\ u=1,2,...,5$	134.381	143.778	158.250	184.095
	119.762	126.851	136.664	150.868
$\hat{\mu}_{y_1}$	135.229	144.964	160.016	187.061
$\hat{\mu}_{y_3}$	133.377	142.384	156.194	180.676
$\hat{\bar{y}}_\tau$	113.018	115.786	119.666	125.657
π_1	110.930	113.197	116.349	121.152
π_3	110.101	112.175	115.049	119.408
$\bar{y}_1$	109.409	111.325	113.972	117.971
$\bar{y}_2$	108.856	110.649	113.118	116.837
$\bar{y}_3$	108.363	110.045	112.359	115.832
$\bar{y}_4$	107.780	109.335	111.467	114.657
$\bar{y}_5$	107.107	108.515	110.442	113.314
$\bar{y}_6$	104.513	105.380	106.553	108.276
$\bar{y}_7$	113.029	116.750	120.994	124.908
$\bar{y}_8$	122.710	128.843	136.911	147.404
$\bar{y}_9$	132.346	141.173	153.989	174.229
$\bar{y}_{MM_1}$	132.346	141.173	153.989	174.229
$\bar{y}_{MM_2}$	138.326	149.558	167.018	198.403
$\bar{y}_{MM_3}$	26.645	23.987	20.998	17.535
$\bar{y}_{MM_4}$	7.695	6.707	5.688	4.610
$\bar{y}_{MM_5}$	4.974	4.314	3.642	2.940
$\hat{\bar{y}}_{\eta_1}$	3.506	3.033	2.554	2.058
$\hat{\bar{y}}_{\eta_2}$	2.667	2.304	1.936	1.559
$\hat{\bar{y}}_{\eta_3}$	2.094	1.806	1.517	1.220
$\hat{\bar{y}}_{\eta_4}$	1.574	1.356	1.138	0.915
$\hat{\bar{y}}_{\eta_5}$	1.129	0.972	0.815	0.654
$\hat{\bar{y}}_{\eta_6}$	0.277	0.238	0.199	0.160
$\hat{\bar{y}}_{\eta_7}$	17.718	15.653	13.490	11.167
$\hat{\bar{y}}_{\eta_8}$	28.052	25.194	22.094	18.635
$\hat{\bar{y}}_{\eta_9}$	16.517	14.565	12.527	10.380
$\bar{y}_{p_1}$	48.875	45.393	41.317	36.357
$\bar{y}_{p_2}$	139.404	151.198	169.732	203.713
$\bar{y}_{p_3}$				
$\bar{y}_{p_4}$				
$\bar{y}^{*}_{p_j},\ j=1,2,3,4$				

where * represents the corrected *PRE* values of estimators due to Al-Hadhrami [11] and Khan *et al.* [19].

Table 4. Effect of skewness and kurtosis over *PREs* of proposed estimators T w.r.t. conventional mean estimator y_m for Gamma population.

Population	Skewness of $y(x)$	Kurtosis of (x)	*PRE(T)*
$x^* \sim Gamma(0.1, 1.0)$ $y^* \sim Gamma(0.85, 2.0)$	4.0432(5.4219)	26.0686(40.0099)	151.5284
$x^* \sim Gamma(0.1, 1.0)$ $y^* \sim Gamma(0.8, 2.0)$	4.1001(5.4219)	26.8685(40.0099)	122.5228
$x^* \sim Gamma(0.1, 1.0)$ $y^* \sim Gamma(0.6, 2.0)$	4.3832(5.4219)	29.9540(40.0099)	119.3913
$x^* \sim Gamma(0.09, 1.0)$ $y^* \sim Gamma(0.5, 2.0)$	4.5960(5.7000)	32.0368(43.8948)	107.9162
$x^* \sim Gamma(0.07, 1.0)$ $y^* \sim Gamma(0.3, 2.0)$	5.0742(6.5113)	37.9816(54.5534)	106.1589
$x^* \sim Gamma(0.06, 1.0)$ $y^* \sim Gamma(0.2, 2.0)$	5.3599(6.7623)	41.7491(57.9522)	103.3111
$x^* \sim Gamma(0.05, 1.0)$ $y^* \sim Gamma(0.3, 2.0)$	5.5988(7.4839)	45.1819(69.9994)	101.7355
$x^* \sim Gamma(0.05, 1.0)$ $y^* \sim Gamma(0.2, 2.0)$	5.7234(7.4839)	46.7809(69.9994)	101.6108
$x^* \sim Gamma(0.04, 1.0)$ $y^* \sim Gamma(0.09, 2.0)$	5.7906(7.4682)	45.6626(67.6872)	101.1449
$x^* \sim Gamma(0.05, 1.0)$ $y^* \sim Gamma(0.1, 2.0)$	6.0014(7.4839)	51.3433(69.9994)	100.0107

Table 5. Effect of skewness and kurtosis over *PREs* of proposed estimators T w.r.t. conventional mean estimator y_m for *Weibull* population.

Population	Skewness of $y(x)$	Kurtosis of (x)	*PRE(T)*
$x^* \sim Wb(1.0, 6.0)$ $y^* \sim Wb(0.7, 1.0)$	1.6991(2.0771)	7.2539(9.4696)	297.4274
$x^* \sim Wb(1.0, 5.9)$ $y^* \sim Wb(0.6, 1.0)$	1.7490(2.0771)	7.4410(9.4396)	293.2841
$x^* \sim Wb(1.0, 5.8)$ $y^* \sim Wb(0.5, 1.0)$	1.8259(2.0771)	7.8021(9.4396)	277.8692
$x^* \sim Wb(1.0, 5.7)$ $y^* \sim Wb(0.4, 1.0)$	1.9584(2.0771)	8.6451(9.4396)	244.7355
$x^* \sim Wb(1.0, 5.6)$ $y^* \sim Wb(0.3, 1.0)$	2.2175(2.0771)	10.9999(9.4396)	189.9646
$x^* \sim Wb(1.0, 5.5)$ $y^* \sim Wb(0.2, 1.0)$	2.7747(2.0771)	18.3644(9.4396)	130.1472
$x^* \sim Wb(1.0, 5.0)$			

(Table 5) cont.....

$y^*{\sim}Wb(0.1, 1.0)$	3.8302(2.0771)	37.0470(9.4396)	105.6356
$x^*{\sim}Wb(1.0, 4.9)$ $y^*{\sim}Wb(0.09, 1.0)$	3.9330(2.0771)	39.0475(9.4396)	105.2101
$x^*{\sim}Wb(1.0, 4.8)$ $y^*{\sim}Wb(0.08, 1.0)$	4.0228(2.0771)	40.8089(9.4396)	104.9641
$x^*{\sim}Wb(1.0, 4.7)$ $y^*{\sim}Wb(0.07, 1.0)$	4.0962(2.0771)	42.2534(9.4396)	104.8430

Table 6. Effect of skewness and kurtosis over *PREs* of proposed estimators *T* w.r.t. conventional mean estimator y_m for *Beta-I* population.

Population	Skewness of $y(x)$	Kurtosis of (x)	*PRE(T)*
$x^*{\sim}Beta(5.5, 7.3)$ $y^*{\sim}Beta(4.3, 6.6)$	0.1001(0.1021)	2.7420(2.7125)	151.5580
$x^*{\sim}Beta(2.5, 4.5)$ $y^*{\sim}Beta(1.5, 2.5)$	0.2175(0.3451)	2.6783(2.6287)	151.1523
$x^*{\sim}Beta(2.5, 4.5)$ $y^*{\sim}Beta(0.5, 0.6)$	0.2479(0.3451)	2.5892(2.6287)	149.0320
$x^*{\sim}Beta(3.0, 5.0)$ $y^*{\sim}Beta(0.5, 1.5)$	0.2568(0.2719)	2.5851(2.6219)	140.6758
$x^*{\sim}Beta(2.0, 4.0)$ $y^*{\sim}Beta(0.5, 1.0)$	0.3082(0.4249)	2.7929(2.6397)	115.3598
$x^*{\sim}Beta(2.0, 4.0)$ $y^*{\sim}Beta(0.5, 1.5)$	0.4357(0.4249)	2.7553(2.6397)	112.9628
$x^*{\sim}Beta(2.0, 4.5)$ $y^*{\sim}Beta(0.5, 2.5)$	0.5041(0.4960)	2.8667(2.7407)	111.2052
$x^*{\sim}Beta(0.5, 1.0)$ $y^*{\sim}Beta(2.0, 4.0)$	0.5874(0.7439)	2.5706(2.3390)	104.4598
$x^*{\sim}Beta(0.5, 1.0)$ $y^*{\sim}Beta(1.0, 2.0)$	0.5958(0.7439)	2.6197(2.3390)	104.0255
$x^*{\sim}Beta(0.7, 3.0)$ $y^*{\sim}Beta(0.9, 2.0)$	1.0091(1.3582)	3.8350(4.5166)	102.8187

Table 7. Effect of skewness and kurtosis over *PREs* of proposed estimators *T* w.r.t. conventional mean estimator y_m for *Chi-square* population.

Population	Skewness of $y(x)$	Kurtosis of (x)	*PRE(T)*
$x^*{\sim}\chi^2(7)$ $y^*{\sim}\chi^2(8)$	0.6510(0.9297)	3.3846(4.2017)	231.3572
$x^*{\sim}\chi^2(7)$ $y^*{\sim}\chi^2(10)$	0.6654(0.9297)	3.5558(4.2017)	202.9646
$x^*{\sim}\chi^2(7)$ $y^*{\sim}\chi^2(12)$	0.7142(0.9297)	3.7801(4.2017)	169.6796
$x^*{\sim}\chi^2(12)$ $y^*{\sim}\chi^2(12)$	0.7522(0.8563)	3.8107(4.1016)	117.8734
$x^*{\sim}\chi^2(8)$ $y^*{\sim}\chi^2(8)$	0.7649(1.0699)	3.7481(4.8351)	104.9040
$x^*{\sim}\chi^2(6)$			

(Table 7) cont.....

$y^*\sim\chi^2(8)$	0.8666(0.9679)	4.1085(4.3849)	103.1570
$x^*\sim\chi^2(8)$ $y^*\sim\chi^2(9)$	0.8695(1.0699)	4.4955(4.8351)	102.3089
$x^*\sim\chi^2(9)$ $y^*\sim\chi^2(8)$	0.8811(1.0606)	4.2927(4.8491)	101.2391
$x^*\sim\chi^2(9)$ $y^*\sim\chi^2(10)$	0.8817(1.0606)	4.5075(4.8491)	100.4092
$x^*\sim\chi^2(9)$ $y^*\sim\chi^2(11)$	0.8975(1.0606)	4.6103(4.8491)	100.4064

APPENDIX I

The *MSE* of the conventional mean estimator under *RSS* is

$$MSE(\bar{y}_m) = \bar{Y}^2\left(\gamma C_y^2 - W_{y_{[i]}}^2\right) \tag{72}$$

The bias and *MSE* of the ratio estimator $\bar{y}_r$ is

$$Bias(\bar{y}_r) = \bar{Y}\left[\left(\gamma C_x^2 - W_{x_{(i)}}^2\right) - \left(\gamma\rho_{xy}\,C_x\,C_y - W_{xy_{[i]}}\right)\right]$$

$$MSE(\bar{y}_r) = \bar{Y}^2\left[\begin{array}{c}\left(\gamma C_y^2 - W_{y_{[i]}}^2\right) + \left(\gamma C_x^2 - W_{x_{(i)}}^2\right) \\ -2\left(\gamma\rho_{xy}\,C_x\,C_y - W_{xy_{[i]}}\right)\end{array}\right] \tag{73}$$

The *MSE* of regression estimator t_β is

$$MSE(t_\beta) = V(\bar{y}_{[n]}) + \hat{\beta}^2 V(\bar{x}_{(n)}) - 2\hat{\beta}Cov(\bar{x}_{(n)}, \bar{y}_{[n]}) \tag{74}$$

The optimum value of $\hat{\beta}$ is obtained by minimizing (74) w.r.t. $\hat{\beta}$

$$\hat{\beta}_{(opt)} = \frac{\hat{R}\left(\gamma\rho_{xy}\,\hat{C}_x\,\hat{C}_y - \widehat{W}_{xy_{[i]}}\right)}{\left(\gamma\hat{C}_x^2 - \widehat{W}_{x_{(i)}}^2\right)}$$

where $\hat{R} = \dfrac{\bar{y}_{[n]}}{\bar{x}_{(n)}}$

The minimum *MSE* at the optimum value of $\hat{\beta}$ is

$$minMSE(t_\beta) = \bar{Y}^2 \left[\left(\gamma C_y^2 - W_{y_{[i]}}^2 \right) - \frac{\left(\gamma \rho_{xy} C_x C_y - W_{xy_{[i]}} \right)^2}{\left(\gamma C_x^2 - W_{x_{(i)}}^2 \right)} \right] \tag{75}$$

The bias and *MSE* of the estimator $\bar{y}_k$ is

$$Bias(\bar{y}_k) = (k-1)\bar{Y} - k\bar{Y} \left[\left(\gamma C_x^2 - W_{x_{(i)}}^2 \right) - \left(\gamma \rho_{xy} C_x C_y - W_{xy_{[i]}} \right) \right]$$

$$MSE(\bar{y}_k) = \bar{Y}^2 \left[\begin{array}{c} (k-1)^2 + k^2 \left(\gamma C_y^2 - W_{y_{[i]}}^2 \right) + \left(\gamma C_x^2 - W_{x_{(i)}}^2 \right) \\ -2k \left(\gamma \rho_{xy} C_x C_y - W_{xy_{[i]}} \right) \end{array} \right] \tag{76}$$

Kadilar *et al.* [10] demonstrated in their study that the estimator proposed by them was more efficient than the estimator due to Prasad [28] proposed under *SRS*.

The optimum value of k is obtained by minimizing (76) w.r.t. k as

$$k_{(opt)} = \frac{\left(1 + \gamma \rho_{xy} C_x C_y - W_{xy_{[i]}} \right)}{\left(1 + \gamma C_y^2 - W_{y_{[i]}}^2 \right)} = k^*(say)$$

The minimum *MSE* at the optimum value of k is

$$minMSE(\bar{y}_k) = \bar{Y}^2 \left[\begin{array}{c} (k^* - 1)^2 + k^{*2} \left(\gamma C_y^2 - W_{y_{[i]}}^2 \right) + \left(\gamma C_x^2 - W_{x_{(i)}}^2 \right) \\ -2k^* \left(\gamma \rho_{xy} C_x C_y - W_{xy_{[i]}} \right) \end{array} \right] \tag{77}$$

The *MSE* of the estimator $\hat{\bar{\mu}}_{y_h}$ is

$$MSE(\hat{\bar{\mu}}_{y_h}) = \bar{Y}^2 \left[(K_h - \beta)^2 \bar{X}^2 \left(\gamma C_x^2 - W_{x_{(i)}}^2 \right) + \gamma S_y^2 (1 - \rho_{xy}^2) \right], h = 1, 3 \tag{78}$$

where $K_h = \dfrac{\bar{Y}}{\bar{X} + q_h}, h = 1,3$

The bias and *MSE* of the estimator $\bar{y}_\tau$ is

$$Bias(\bar{y}_\tau) = (\tau - 1)\bar{Y} - \tau \bar{Y} \left(\gamma C_x^2 - W_{x_{(i)}}^2 \right) - \left(\gamma \rho_{xy} C_x C_y - W_{xy_{[i]}} \right)$$

$$MSE(\bar{y}_\tau) = \begin{bmatrix} \bar{Y}^2(1-2\tau)^2 + \bar{Y}^2\tau^2\left(\gamma C_y^2 - W_{y_{[i]}}^2\right) + \left(\gamma C_x^2 - W_{x_{(i)}}^2\right) \\ -2\tau\left(\gamma\rho_{xy}\,C_x\,C_y - W_{xy_{[i]}}\right) \end{bmatrix} \tag{79}$$

The *MSE* of the estimator Π_h is

$$MSE(\Pi_h) = \bar{Y}^2\begin{bmatrix} \bar{Y}^2\left(\gamma C_y^2 - W_{y_{[i]}}^2\right) + I_h^2\bar{X}^2\left(\gamma C_x^2 - W_{x_{(i)}}^2\right) \\ -2I_h\bar{X}\bar{Y}\left(\gamma\rho_{xy}\,C_x\,C_y - W_{xy_{[i]}}\right) \end{bmatrix} \tag{80}$$

where $I_h = 2\bar{Y}\left(\frac{\bar{X}+q_h}{(2\bar{X}+q_h)^2}\right)$

The *MSE* of the estimator $\bar{y}_l$ is

$$MSE(\bar{y}_l) = \bar{Y}^2\left[(Z_l - \beta)^2\bar{X}^2\left(\gamma C_x^2 - W_{x_{(i)}}^2\right) + \gamma S_y^2(1 - \rho_{xy}^2)\right] \tag{81}$$

where $Z_l = \left(\frac{\bar{Y}}{(\bar{X}+D_l)}\right)\beta_1(x)$, $l = 1,2,\dots,9$

The biases and *MSEs* of the estimator $\bar{y}_{MM_j}$, $j = 1,2,\dots,5$ are

$$Bias\left(\bar{y}_{MM_j}\right) = \bar{Y}\begin{bmatrix} \gamma_j^2\left(\gamma C_x^2 - W_{x_{(i)}}^2\right) \\ -\gamma_j\left(\gamma\rho_{xy}\,C_x\,C_y - W_{xy_{[i]}}\right) \end{bmatrix},\, j = 1,2,3$$

$$Bias(\bar{y}_{MM_4}) = \bar{Y}\left[\gamma_4\left(\gamma\rho_{xy}\,C_x\,C_y - W_{xy_{[i]}}\right)\right]$$

$$Bias(\bar{y}_{MM_5}) = \bar{Y}\begin{bmatrix} \gamma t_3 C_x^2(k + \alpha t_3(t_3 + 2K)) \\ -(\alpha t_3^2 W_{x_{(i)}}^2 + (1 - 2\alpha)t_3 W_{xy_{[i]}}) \end{bmatrix}$$

and

$$MSE\left(\bar{y}_{MM_j}\right) = \bar{Y}^2\begin{bmatrix} \left(\gamma C_y^2 - W_{y_{[i]}}^2\right) + \gamma_j^2\left(\gamma C_x^2 - W_{x_{(i)}}^2\right) \\ -2\gamma_j\left(\gamma\rho_{xy}\,C_x\,C_y - W_{xy_{[i]}}\right) \end{bmatrix},\, j = 1,2,3 \tag{82}$$

$$MSE(\bar{y}_{MM_4}) = \bar{Y}^2 \left[\begin{array}{c} \left(\gamma C_y^2 - W_{y_{[i]}}^2\right) + \gamma_4^2 \left(\gamma C_x^2 - W_{x_{(i)}}^2\right) \\ +2\gamma_4 \left(\gamma \rho_{xy} \, C_x \, C_y - W_{xy_{[i]}}\right) \end{array} \right] \tag{83}$$

$$MSE(\bar{y}_{MM_5}) = \bar{Y}^2 \left[\begin{array}{c} \gamma\{C_y^2 + (1-2\alpha)t_3 C_x^2 \left((1-2\alpha)t_3 + 2K\right)\} \\ -\left\{W_{y_{[i]}} - (1-2\alpha)t_3 W_{x_{(i)}}\right\}^2 \end{array} \right] \tag{84}$$

The optimum value of α is obtained by minimizing (84) w.r.t. α as

$$\alpha_{(opt)} = \frac{(t_3 + K)}{2t_3}$$

The minimum *MSE* at optimum value of α is

$$MSE(\bar{y}_{MM_5}) = \bar{Y}^2 \left[\gamma S_y^2 (1 - \rho_{xy}^2) - \left(W_{y_{[i]}} - \frac{(t_3 - K)}{2} W_{x_{(i)}}\right)^2 \right] \tag{85}$$

where $\gamma = 1/mr$, $\gamma_1 = \bar{X}/(\bar{X} + C_x)$, $\gamma_2 = \bar{X}/(\bar{X} + \beta_2(x))$, $\gamma_3 = \gamma_4 = t_3 = \bar{X}C_x/(\bar{X}C_x + \beta_2(x))$, $K = \rho_{xy}(C_y/C_x)$

The *MSE* of the estimator $\bar{y}_{\eta_l}$ is

$$MSE(\bar{y}_{\eta_l}) = \bar{Y}^2 \left[\begin{array}{c} \left(\gamma C_y^2 - W_{y_{[i]}}^2\right) + \eta_l^2 \left(\gamma C_x^2 - W_{x_{(i)}}^2\right) \\ -2\eta_l \left(\gamma \rho_{xy} \, C_x \, C_y - W_{xy_{[i]}}\right) \end{array} \right], l = 1, 2, \ldots, 9 \tag{86}$$

CONSENT FOR PUBLICATION

Not applicable.

CONFLICT OF INTEREST

There is no conflict of interest declared.

ACKNOWLEDGEMENTS

The authors would like to express their gratitude to the Director Publications, Bentham Science Publishers Pte. Ltd., for his valuable suggestions.

REFERENCES

[1]   G.A. McIntyre, "A method of unbiased selective sampling using ranked set", *Australian J. Agricultural Research,* vol. 3, pp. 385-390, 1952.
[http://dx.doi.org/10.1071/AR9520385]

[2]   T.R. Dell, *"The theory and some applications of ranked set sampling Doctoral dissertation",* University of Georgia, Athens, GA, 1969.

[3]   T.R. Dell, and J.L. Clutter, "Ranked set sampling theory with order statistics background", *Biometrics,* vol. 28, pp. 545-555, 1972.
[http://dx.doi.org/10.2307/2556166]

[4]   K. Takahasi, and K. Wakimoto, "On unbiased estimates of the population mean based on the sample stratified by means of ordering", *Annals of the Institute of Statistical Mathematics,* vol. 20, pp. 1-31, 1968.
[http://dx.doi.org/10.1007/BF02911622]

[5]   L.S. Halls, and T.R. Dell, "Trial of ranked set sampling for forage yields", *Forest Sci.,* vol. 12, no. 1, pp. 22-26, 1966.

[6]   H.A. Muttlak, and L.L. McDonald, "Ranked set sampling with respect to a concomitant variables and with size biased probability of selection," *Communications in Statistics - Theory and Methods,* vol. 19, no. l, pp. 205-219, 1990.
[http://dx.doi.org/10.1080/03610929008830198]

[7]   H.A. Muttlak, and L.L. McDonald, "Ranked set sampling and line intercept method: A more efficient procedure", *Biometric Journal,* vol. 34, pp. 329-346, 1992.
[http://dx.doi.org/10.1002/bimj.4710340307]

[8]   H.A. Muttlak, "Parameter estimation in simple linear regression using ranked set sampling", *Biometric Journal,* vol. 37, no. 7, pp. 799-810, 1995.
[http://dx.doi.org/10.1002/bimj.4710370704]

[9]   H.M. Samawi, and H.A. Muttlak, "Estimation of ratio using ranked set sampling", *Biometric Journal,* vol. 38, pp. 753-764, 1996.
[http://dx.doi.org/10.1002/bimj.4710380616]

[10]   C. Kadilar, Y. Unyazici, and H. Cingi, "Ratio estimator for the population mean using ranked set sampling", *Stat Papers,* vol. 50, pp. 301-309, 2007.

[11]   S.A. Al-Hadhrami, "Ratio type estimators of the population mean based on ranked set sampling", *International Journal of Mathematical, Computational, Physical, Electrical and Computer Engineering,* vol. 3, no. 11, pp. 896-900, 2009.

[12]   C. Kadilar, and H. Cingi, "Ratio estimators in simple random sampling", *Applied Mathematics and Computation,* vol. 151, pp. 893-902, 2004.
[http://dx.doi.org/10.1016/S0096-3003(03)00803-8]

[13]   A.I. Al-Omari, A.A. Jemain, and K. Ibrahim, "New ratio estimators of the mean using simple random sampling and ranked set sampling methods", *Revista Investigation Operational,* vol. 30, no. 2, pp. 97-108, 2009.

[14]   M.I. Jeelani, and C.N. Bouza, "New ratio method of estimation under ranked set sampling", *Revista Investigation Operational,* vol. 36, no. 2, pp. 151-155, 2015.

[15]   M.I. Jeelani, S.E.H. Rizvi, M.K. Sharma, S.A. Mir, T.A. Raja, N. Nazir, and F. Jeelani, "Improved ratio estimation under ranked set sampling", *International Journal of Modern Mathematical Sciences,* vol. 14, no. 2, pp. 204-211, 2016.

[16]   M. Saini, and A. Kumar, "Ratio estimators for the finite population mean under simple random sampling and ranked set sampling", *Int Jour Syst Assur Eng Manag,* vol. 8, no. 2, pp. 488-492, 2016.
[http://dx.doi.org/10.1007/s13198-016-0454-y]

[17]   N. Mehta, and V.L. Mandowara, "A modified ratio-cum-product estimator of finite population mean using ranked set sampling", *Communications in Statistics - Theory and Methods,* vol. 45, no. 2, pp. 267-276, 2016.

[http://dx.doi.org/10.1080/03610926.2013.830748]

[18]   M.I. Jeelani, C.N. Bouza, and M. Sharma, "Modified ratio estimator under ranked set sampling", *Revista Investigation Operational,* vol. 38, no. 1, pp. 103-106, 2017.

[19]   L. Khan, J. Shabbir, and A. Khalil, *A new class of regression cum ratio estimators of population mean in ranked set sampling.* Life Cycle Reliability and Safety Engineering, 2019, pp. 1-4.

[20]   S. Bhushan, and A. Kumar, "On optimal classes of estimators under ranked set sampling", *Communications in Statistics - Theory and Methods,* 2020.
[https://doi.org/10.1080/03610926.2020.1777431]

[21]   S. Bhushan, R. Gupta, and S.K. Pandey, "Some log-type classes of estimators using auxiliary information", *International Journal of Agricultural and Statistical Sciences,* vol. 11, no. 2, pp. 487-491, 2015.

[22]   S. Bhushan, and R. Gupta, "Some log-type classes of estimators using auxiliary attribute", *Advances in Computational Sciences and Technology,* vol. 12, no. 2, pp. 99-108, 2019. a

[23]   S. Bhushan, and R. Gupta, "An improved log-type family of estimators using attribute", *Journal of Statistics and Management System,* 2019. b
[http://dx.doi.org/10.1080/09720510.1661604]

[24]   S. Bhushan, and R. Gupta, "Some new log-type class of double sampling estimators", *International Journal of Applied Agricultural Research,* vol. 14, no. 1, pp. 31-40, 2019. c

[25]   S. Bhushan, and R. Gupta, "A class of log-type estimators for population mean using auxiliary information on an attribute and a variable using double sampling technique", *International Journal of Computational and Applied Mathematics,* vol. 14, no. 1, pp. 1-10, 2019. d

[26]   S. Bhushan, and R. Gupta, "Searls' ratio product type estimators", *International Journal of Statistics and System,* vol. 14, no. 1, pp. 29-37, 2019. e

[27]   P.L.H. Yu, and K. Lam, "Regression estimator in ranked set sampling", *Biometrics,* vol. 53, no. 3, pp. 1070-1080, 1997.
[PMID: 9333340]

[28]   B. Prasad, "Some improved ratio type estimators of population mean and ratio in finite population sample surveys", *Communications in Statistics - Theory and Methods,* vol. 18, pp. 379-392, 1989.

[29]   B.V.S. Sisodia, and V.K. Dwivedi, "A modified ratio estimator using coefficient of variation of auxiliary variable", *Journal of Indian Society of Agricultural Statistics,* vol. 33, pp. 13-18, 1981.

[30]   H.P. Singh, and M.S. Kakran, "A modified ratio estimator using known coefficient of kurtosis of an auxiliary character", In: S Singh, Ed., *Advanced Sampling Theory with Applications,* vol. 2. Kluwer Academic Publishers, 1993.

[31]   L.N. Upadhyaya, and H.P. Singh, "Use of transformed auxiliary variable in estimating the finite population mean", *Biometrical Journal,* vol. 41, pp. 627-636, 1999.

[32]   H.P. Singh, and M.R. Espejo, "On linear regression and ratio estimator using coefficient of variation of auxiliary variate", *Statistician,* vol. 52, no. 1, pp. 59-67, 2003.
[http://dx.doi.org/10.1111/1467-9884.00341]

[33]   H.P. Singh, and S. Horn, "An alternative estimator for multi-character surveys", *Metrika,* vol. 48, pp. 99-107, 1998.

CHAPTER 5

Analysis of Bivariate Survival Data using Shared Inverse Gaussian Frailty Models: A Bayesian Approach

Arvind Pandey[1], Shashi Bhushan[2], Lalpawimawha[3,*] and Shikhar Tyagi[1]

[1]*Department of Statistics, Central University of Rajasthan, India*
[2]*Department of Mathematics and Statistics, Dr. Shakuntala Misra National Rehabilitation University, Lucknow, India*
[3]*Department of Statistics, Pachhunga University College, Mizoram, India*

Abstract: Frailty models are used in the survival analysis to account for the unobserved heterogeneity in individual risks of disease and death. The shared frailty models have been suggested to analyze the bivariate data on related survival times (*e.g.*, matched pairs experiments, twin or family data). This paper introduces the shared Inverse Gaussian (IG) frailty model with baseline distribution as Weibull exponential, Lomax, and Logistic exponential. We introduce the Bayesian estimation procedure using Markov Chain Monte Carlo (MCMC) technique to estimate the parameters involved in these models. We present a simulation study to compare the actual values of the parameters with the estimated values. Also, we apply these models to a real-life bivariate survival data set of McGilchrist and Aisbett [1] related to the kidney infection data, and a better model is suggested for the data.

Keywords: Bayesian model comparison, Inverse gaussian frailty, Lomax distribution, Logistic exponential distribution, MCMC, Shared frailty, Weibull exponential distribution.

INTRODUCTION

The statistical analysis of time-to-event, event-history, or duration data plays an essential role in medicine, epidemiology, biology, demography, engineering, actuarial science, and other fields. In the past several years, medical research concerning the addition of random effects to the survival model has substantially increased. The random effect model is a model with a continuous random variable presenting excess risk or frailty for individuals or families. Sometimes due to the

---

*Corresponding author Lalpawimawha: Department of Statistics, Pachhunga University College, Mizoram, India; Tel: +91-9862307640; E-mail: raltelalpawimawha08@gmail.com.

economic reasons or human ignorance or non-availability of the factors, the crucial factors are unobserved in the model. This unobserved factor is usually termed as heterogeneity or frailty. In the statistical modeling concept, the frailty approach covers the heterogeneity caused by unmeasured covariates like genetic factors or environmental factors. In the frailty model, the random effect (frailty) has a multiplicative effect on the baseline hazard function, extending the Cox proportional hazard model. Clayton [2] introduced a random effect model to account for the frailty shared by all individuals in a group.

Generally, this heterogeneity is often referred to as variability in survival analysis and is considered one of the critical sources of variability in medical and biological applications. It may not be easy to assess, but it is nevertheless of great importance in the model. In the model, if the frailty is either statistically impactful or ignored, the model will be unsuitable, and the decision based on such models will be misleading; ignorance of the frailty may lead to either underestimation or overestimation of the parameters and higher values of AIC, BIC, DIC can be seen in comparison to the model (after including frailty) [3]. In general terms, we let the heterogeneity go into the error term. It leads to an increase in the response's variability compared to the case when frailty is included. The unobservable risks are random variables, which follow some distribution. It is possible to choose a different distribution for unobserved covariates. The variance of the frailty distribution determines the degree of heterogeneity in the study population. This paper considers the shared frailty model with a random effect or frailty in the hazard model, common and shared by all individuals in the group [3, 4]. Since the frailty is not observed, we assume to follow a positive stable distribution.

In practice, the gamma frailty specification may not fit well [5-7]. However, gamma frailty has drawbacks. For example, it may weaken the effect of covariates studied by Hougaard [8] in the analysis of multivariate survival data. IG distribution can be another practical choice. Concerning time, the population becomes homogeneous under IG, whereas the relative heterogeneity remains constant for gamma [9], udder quarter infection data; Duchateau and Janssen [10] fit the IG frailty model with the Weibull hazard rate.

The gamma model has predictive hazard ratios that are time-invariant and may not be suitable for these patterns of failures [11].

The term "frailty" itself was first introduced by Vaupel *et al.* [4] in univariate survival models and was substantially promoted by its applications to the multivariate survival data. In shared frailty models, we assumed that the survival times are conditionally independent for the given shared frailty. That means dependence between survival times is only due to frailty. In the frailty model, the unobserved random effect acts multiplicatively on baseline hazard function, which is assumed to follow one of the parametric distributions like gamma, IG, positive stable, log-normal, and power variance function. Let T be a continuous lifetime random variable, and random variable Z be a frailty variable. The conditional hazard function for given frailty at $t > 0$ is given by

$$h(t|z)=zh_0(t)e^{X\beta} \tag{1.1}$$

where $h_0(t)$ is a baseline hazard function at time $t > 0$. X is a row vector of covariates, and β is a column vector of regression coefficients. The conditional survival function for given frailty at time $t > 0$ is,

$$S(t|z)=e^{-\int_0^t h_0(x|z)dx} \tag{1.2}$$

$$=e^{-zH_0(t)e^{X\beta}} \tag{1.3}$$

Where $H_0(t)$ is the cumulative baseline hazard function at time $t > 0$. Integrating over the range of frailty variable Z having density $f(z)$, we get the marginal survival function as,

$$S(t)=\int_0^\infty S(t|z)f(z)dz$$

$$=\int_0^\infty e^{-zH_0(t)e^{X\beta}}f(z)dz \tag{1.4}$$

$$=L_Z(H_0(t)e^{X\beta}) \tag{1.5}$$

where $L_Z (.)$ is the Laplace transform of the distribution of Z. Once we get the survival function at time $t > 0$ of lifetime random variable for an individual, we can obtain probability structure and the inference based on it.

The remaining article is organized as follows. In Sec. 2, the introduction of the generally shared frailty model is provided; also, we have discussed IG shared frailty models, respectively. In Sec. 3 and 4, we have introduced a shared frailty model and baseline distributions. Different proposed shared frailty models are given in section 5. An outline of model fitting, using the Bayesian approach, is presented in Sec. 6.

Secs. 7 and 8 are devoted to the simulation study and kidney infection data analysis, respectively. Finally, in Sec. 9, we have discussed the results.

GENERAL SHARED FRAILTY MODEL

The shared frailty model is a conditional independence model where the frailty is common to all individuals in a cluster and responsible for creating dependence between events. It is relevant to the event-time of related individuals, the pair of organs, and repeated measurements. If the failure of paired organs like kidneys, lungs, eyes, ears, and dental implants is considered, it is assumed that all failure times in a cluster are conditionally independent of the frailties. In this model, the individuals from the same cluster share common covariates, and the value of the frailty term is constant over time. If the variation of the frailty variable Z is zero, this implies independence between event times in the clusters; otherwise, there is positive dependence between event times within the cluster.

The conditional hazard model for i^{th} cluster at j^{th}, $(j=1,2)$ survival time $t_{ij} > 0$, for given frailty $Z_i = z_i$ and observed covariate vector X_i is:

$$h(t_{ij}|Z_i, X_i) = z_i h_0(t_{ij})\, e^{X'\beta} \tag{2.1}$$

where Z_i refers to random multiplicative effects or "frailties, shared by all members of the same cluster, $h_0(t_{ij})$ is the common baseline hazard function, and β is a vector of unknown regression coefficients.

Model (2.1) is called the shared frailty model because individuals in the same cluster share the same frailty factor.

Under the assumption of independence, the conditional survival function in the bivariate case for given frailty $Z_i = z_i$ at time $t_{i1} > 0$ and $t_{i2} > 0$ is

$$S(t_{i1}, t_{i2}|Z_i, X_i) = S(t_{i1}|Z_i, X_i)\, S(t_{i2}|Z_i, X_i)$$

$$= e^{-z_i(H_{01}(t_{i1})+H_{02}(t_{i2}))} e^{x_i'\beta} \; ; \; i=1,2\ldots, n \; ; \; j=1,2. \tag{2.2}$$

where $H_0(t_{ij})$ is the cumulative baseline hazard function at time $t_{ij} > 0$.

The unconditional bivariate survival function at time $t_{i1} > 0$ and $t_{i2} > 0$ can be obtained by integrating out the frailty variable Z_i having the probability function $f(z_i)$, for the ith individual.

$$S(t_{i1}, t_{i2}|X_i) = \int_{Z_i} S(t_{i1}, t_{i2}|z_i)\, f(z_i)\, dz_i$$

$$= L_Z[(H_{01}(t_{i1}) + H_{02}(t_{i2}))\, e^{x_i'\beta}] \tag{2.3}$$

Where L_{Z_i} is the Laplace transform of the distribution of Z_i. Thus, the bivariate survivor function is easily expressed using the Laplace transform of the frailty distribution, evaluated at the total integrated conditional hazard.

INVERSE GAUSSIAN FRAILTY

In shared frailty models, the most commonly used frailty distribution is gamma distribution because of its mathematical convenience. Hougaard [9] introduced IG distribution as an alternative to gamma distribution when early occurrences of failures are dominant in a lifetime distribution, and its failure rate is expected to be non-monotonic. The IG distribution has many similarities to standard Gaussian distribution [12]. The early occurrences of failures and expected non-monotonic failure rate provide many stances in modeling. For lifetime models, IG distribution may provide a valuable choice under such situations. Concerning time, the population becomes homogeneous under IG. In contrast, the relative heterogeneity remains constant for gamma, [9] remarked that survival models with gamma and inverse Gaussian frailties behave very differently, noting that the relative frailty distribution among survivors is independent of age for the gamma, but becomes more homogeneous with time for the IG.

To counter the identifiability, assume E[Z]=1. Due to restriction, density function and Laplace transformation of IG distribution reduces to,

$$f(z) = \begin{cases} \left[\dfrac{1}{2\pi\theta}\right]^{\frac{1}{2}} z^{-\frac{3}{2}} e^{\frac{(z-1)^2}{2z\theta}} & ; z>0, \theta>0 \\ 0 & ; Otherwise \end{cases} \tag{3.1}$$

and the Laplace transform is

$$L_z(s) = \exp\left[\frac{1-(1+2\theta)^{1/2}}{\theta}\right] \tag{3.2}$$

with the variance of Z is θ. The frailty variable Z is degenerate at $Z=1$ when θ tends to zero. Note that there is heterogeneity if $\theta > 0$.

Replacing the Laplace transformation in equation (2.3), we get the unconditional bivariate survival function for i^{th} cluster at time $t_{i1} > 0$ and $t_{i2} > 0$ is,

$$S(t_{1j}, t_{2j}) = \exp\left[\frac{1 - (1 + 2\theta e^{x\beta}((H_{01}(t_{i1}) + H_{02}(t_{i2})))^{1/2}}{\theta}\right]$$

Where $H_{01}(t_{i1})$ and $H_{02}(t_{i2})$ are the cumulative baseline hazard functions of the lifetime random variables T_{i1} and T_{i2}, respectively.

BASELINE DISTRIBUTIONS

Weibull distribution and exponential distribution are the most commonly used baseline distribution in survival analysis. Both are flexible distributions that can encompass the characteristics of several other distributions. This property has given rise to widespread applications. Oguntunde, P.E. *et al.* [13] introduced another alternative distribution called Weibull exponential distribution.

If a continuous random variable T follows Weibull exponential distribution, then the survival function, hazard function, and cumulative hazard function are respectively;

$$S(t) = e^{-\alpha(e^{\lambda t}-1)^{\gamma}} \quad \alpha, \lambda, \gamma > 0, \, t \geq 0 \tag{4.1}$$

$$h(t) = \frac{\alpha\lambda\gamma(1-e^{-\lambda t})^{\gamma-1}e^{\lambda\gamma t - \alpha(e^{\lambda t}-1)^{\gamma}}}{e^{-\alpha(e^{\lambda t}-1)^{\gamma}}} \tag{4.2}$$

$$H(t) = \alpha(e^{\lambda t} - 1)^{\gamma} \tag{4.3}$$

If $\gamma = 1$ and $\alpha = \theta/\lambda$, for $\theta > 0$, the Weibull exponential distribution reduces to 0 give the Gompertz distribution [14], and if $\gamma = 1$, the Weibull exponential distribution would reduce to give the exponential distribution. The shape of Weibull exponential distribution is unimodal or decreasing (depending on the parameters). The Weibull exponential distribution is useful as a life testing model.

The second baseline distribution we have used is the Lomax distribution. A continuous random variable T is said to follow if its survival function is,

$$S(t) = (1 + \lambda t)^{-\alpha}, \quad \alpha > 0, \, \lambda > 0, \, t \geq 0 \tag{4.4}$$

Where λ and α are respectively, the scale and shape parameters of the distribution. The hazard function and cumulative hazard function are respectively,

$$h(t) = \frac{\alpha\lambda}{1+\lambda t}, \, t \geq 0 \tag{4.5}$$

$$H(t) = \alpha \ln(1+\lambda t), \, t \geq 0 \tag{4.6}$$

It has monotone failure rate behavior depending on the value of shape parameter α.

The third baseline distribution we have considered is the Logistic exponential distribution. A continuous random variable is said to follow Logistic exponential distribution if its survival function is,

$$S(t) = \frac{\alpha\lambda(e^{\lambda t}-1)^{\lambda-1}e^{\alpha t}}{1+(e^{\alpha t}-1)^{\lambda}}, \, t \geq 0 \tag{4.7}$$

Where λ and α are respectively, scale and shape parameters of the distribution. The hazard function and the cumulative hazard function are respectively,

$$h(t) = \frac{\alpha\lambda(e^{\lambda t}-1)^{\lambda-1}e^{\alpha t}}{1+(e^{\alpha t}-1)^{\lambda}}, \, t \geq 0 \tag{4.8}$$

$$H(t) = \ln(1 + (e^{\alpha t-1})^{\lambda}), \, t \geq 0 \tag{4.9}$$

The distribution reduces to an exponential distribution when $\lambda = 1$, which belongs to both increasing failure rate and decreasing failure rate classes. The distribution is in the Bathtub-Shaped failure rate class when $0 < \lambda < 1$ and in the upside-down bathtub-shaped failure rate class when $\lambda > 1$.

PROPOSED MODELS

Substituting the cumulative hazard function for the Weibull exponential, Lomax and Logistic exponential distributions in Eq. (3.3), we get the unconditional bivariate survival functions at time t_{i1} and t_{i2} as,

$$S(t_{1j}, t_{2j}) = \exp\left[\frac{1-(1+2\theta e^{x\beta}((\alpha_1(e^{\lambda_1 t_{i1}}-1)^{\gamma_1}+\alpha_2(e^{\lambda_2 t_{i2}}-1)^{\gamma_2}))^{1/2}}{\theta}\right] \tag{5.1}$$

$$S(t_{1j}, t_{2j}) = \exp\left[\frac{1 - (1 + 2\theta e^{x\beta}((\alpha_1 \ln(1 + \lambda_1 t_{i1}) + \alpha_2 \ln(1 + \lambda_2 t_{i2})))^{1/2}}{\theta}\right] \qquad (5.2)$$

$$S(t_{1j}, t_{2j}) = \exp\left[\frac{1 - (1 + 2\theta e^{x\beta}((\ln(1 + (e^{\alpha_1 t_{i1}} - 1)^{\lambda_1}) + \ln(1 + (e^{\alpha_2 t_{i2}} - 1)^{\lambda_2})))^{1/2}}{\theta}\right] \qquad (5.3)$$

Here onwards, we call equations (5.1), (5.2), and (5.3) as Model I, Model II, and Model III, respectively.

BAYESIAN ESTIMATION OF PARAMETERS AND MODEL COMPARISONS

N individuals are considered understudy, suppose (t_{i1}, t_{i2}) stand for first and second observed failure times. Correspondingly, for first and second recurrence times, c_{i1} and c_{i2} suppose to be the observed censoring times for the ith individual ($i = 1, 2, 3,...,n$). Independence between censoring schemes and lifetimes of individuals has been assumed. The contribution of bivariate lifetime random variable of the ith individual in likelihood function is given by,

$$L_i(t_{i1}, t_{i2}) = \begin{cases} f_1(t_{i1}, t_{i2}), \; ; t_{i1} < c_{i1}, t_{i2} < c_{i2}, \\ f_2(t_{i1}, c_{i2}), \; ; t_{i1} < c_{i1}, t_{i2} > c_{i2}, \\ f_3(c_{i1}, t_{i2}), \; ; t_{i1} > c_{i1}, t_{i2} < c_{i2}, \\ f_4(c_{i1}, c_{i2}), \; ; t_{i1} > c_{i1}, t_{i2} > c_{i2}. \end{cases}$$

and the likelihood function is,

$$L(\psi, \beta, \theta) = \prod_{i=1}^{n_1} f_1(t_{i1}, t_{i2}) \prod_{i=1}^{n_2} f_2(t_{i1}, c_{i2}) \prod_{i=1}^{n_3} f_3(c_{i1}, t_{i2}) \prod_{i=1}^{n_4} f_4(c_{i1}, c_{i2}) \qquad (6.1)$$

where θ, ψ, and β are the frailty parameter, the vector of baseline parameters, and the regression coefficients vector. Let n_1, n_2, n_3 and n_4 be the number of pairs for which first and second failure times (t_{i1}, t_{i2}) lie in the ranges $t_{i1} < c_{i1}, t_{i2} < c_{i2}$; $t_{i1} < c_{i1}, t_{i2} > c_{i2}$; $t_{i1} > c_{i1}, t_{i2} < c_{i2}$ and $t_{i1} > c_{i1}, t_{i2} > c_{i2}$ respectively and

$$f_1(t_{i1}, t_{i2}) = \frac{\partial^2 (t_{i1}, t_{i2})}{\partial t_{i1} \partial t_{i2}}$$

$$f_2(t_{i1}, c_{i2}) = -\frac{\partial (t_{i1}, w_{i2})}{\partial t_{i1}}$$

$$f_3(c_{i1}, t_{i2}) = \frac{\partial (c_{i1}, t_{i2})}{\partial t_{i2}}$$

$$\text{and} \quad f_4(c_{i1}, c_{i2}) = S(c_{i1}, c_{i2})$$

(6.2)

Substituting the hazard function $h_{01}(t_{i1})$, $h_{02}(t_{i2})$, and survival function $S(t_{i1}, t_{i2})$ for the three proposed models, we get the likelihood function given by equation (6.1).

The likelihood equations obtained from the likelihood function (6.1) are not easy to solve. Hence, we have to use the Newton-Raphson iterative procedure to estimate the models involved, but due to many parameters, MLEs do not converge. So we move to the computational Bayesian approach, which does not suffer from these difficulties.

The joint posterior density function of parameters for given failure times is given as,

$$\pi(\alpha_1, \lambda_1, \gamma_1, \alpha_2, \lambda_2, \gamma_2, \theta, \beta) \propto L(\alpha_1, \lambda_1, \gamma_1, \alpha_2, \lambda_2, \gamma_2, \theta, \beta)$$

$$\times$$

$$g_1(\alpha_1) g_2(\lambda_1) g_3(\gamma_1) g_4(\alpha_2) g_5(\lambda_2) g_6(\gamma_2) g_7(\theta) \prod_{i=1}^{5} p_i(\underline{\beta}_{-i})$$

where baseline parameters are supposed to follow prior density functions $g_i(.)$ ($i = 1, 2, \cdots, 5$) with known hyperparameters and frailty variance, respectively. Consequently, regression coefficient β_i having $p_i(.)$ prior density function; $\underline{\beta}_i$ represents a vector of regression coefficients except *for* β_i, $i = 1, 2, \ldots k$, and equation (6.1) used to obtain likelihood function $L(.)$. Independence between parameters has also been assumed. The prior distributions have to be assumed as flat because any prior information about baseline parameters α_1, λ_1, α_2, and λ_2 is not available.

Due to baseline parameters, gamma distribution becomes a widely used non-informative prior with mean and large variance. $G(\varphi, \varphi)$, say with $\varphi = 0.0001$. We consider another non-informative prior $U(b_1, b_2)$, say with $b_1 = 0$ and $b_2 = 100$. Regression coefficients are supposed to follow a normal distribution as prior with mean zero, and large variance, say $s^2 = 1000$. Similar types of prior were used in [15-17].

Metropolis-Hastings Algorithm has been used to fit all the proposed models with the above-mentioned prior density function and likelihood function (6.1). With the different staring points, two chains have been generated. Gelman-Rubin convergence statistic and Geweke test have been used to monitor the Markov chain's convergence to a stationary distribution. Trace plots, coupling from the past plots, and sample autocorrelation function plots are used to check the chain's behavior to decide the burn-in-period and autocorrelation lag.

The predictive intervals have been used to check the models' adequacy by generating samples from the posterior predictive density. Bayesian Information Criteria (BIC), Akaike Information Criteria (AIC), and Deviance Information Criteria (DIC) are Bayesian model selection criteria that have been used to compare the proposed models. Also, we have used Bayes factor B_{uv} for comparison of the model's M_u against M_v. To compute the Bayes factor, we have considered the MCMC approach given previously [18].

SIMULATION STUDY

A simulation study has been done to evaluate the performance of the Bayesian estimation procedure. Only one covariate $\underline{X} = X_1$ that follows normal distribution has been considered for the simulation purpose. As the Bayesian methods are time-consuming, we generate only fifty lifetimes using the inverse transform technique. Estimates of parameters are nearly the same because both the chains were showing somewhat similar results. Therefore, there is no effect of the prior distribution on posterior summaries, so we present the analysis for only one chain with $G(a_1, a_2)$ as before baseline parameters for all the models. Gelman-Rubin convergence statistic values are somehow equivalent to one, and Geweke test values are relatively small, and consequently, p-values are large enough to say the chain attains stationary distribution for both the initial sets. Also, the convergence rate was not significantly different. Tables **1-3** present estimates, credible intervals, Gelman-Rubin convergence statistic, and Geweke test for all the models I, II, and III based on simulation study for the parameters.

ANALYSIS OF KIDNEY INFECTION DATA

The Bayesian procedure is applied to kidney infection data [1]. This data consists of 38 patients, recurrence times (in days) of infection are given, which can be outlined as these are recorded from the catheter's insertion until it must be removed after infection (Needs elaboration). Data include five known covariates age, sex, and disease type Glomerulo Neptiritis (GN), Acute Neptirtis (AN), and Polycystic Kidney Disease (PKD). The catheter may have to be removed for reasons other than kidney infection, which is regarded as censoring. So survival time for a given patient may be first or second infection time or censoring time. After the occurrence or censoring of the first infection, sufficient (ten weeks interval) time was allowed for the infection to be cured before the second time the catheter was inserted. So the first and second recurrence times are taken to be independent apart from the common frailty component.

Opine first, and T1 and T2 symbolize a second time to infection. Five covariates, age, sex, GN, AN, and PKD, are symbolized by X_1, X_2, X_3, X_4, and X_5. Here, the Kolmogorov-Smirnov (K-S) test has been applied to check the goodness of fit of the data set for IG frailty distribution with Weibull exponential, Lomax, and Logistic exponential baseline distributions and then apply the Bayesian estimation procedure, and then further estimation procedure has been applied. Consequently, the p-values of the goodness of fit test for Model I, Model II, and Model III are correspondingly pointed out in Table **4**. Thus, the hypothesis that data are from Weibull exponential, Lomax, and Logistic exponential distributions cannot be rejected based on the K-S test's p values.

Under the model I, 76, 75, 72, 69 and 55 out of 76 observations were included in the 99%, 95%, 90%, 75%, and 50% predictive intervals. While under model II and model III, respectively, the number of observations was 75, 64, 60, 48, and 36; 76, 75, 75, 67, and 55. All three models may seem acceptable for kidney infection data based on observations under predictive intervals. As in the case of simulation, the same set of prior distributions has been assumed. Under the Bayesian estimation procedure, two parallel chains have been run. Two sets of prior distributions have been used with different starting points using the Metropolis-Hastings algorithm and Gibbs sampler based on normal transition kernels. Both chains have been iterated 100000 times. It can be said that estimates are independent of the different prior distributions because, for both sets of priors, estimates of parameters are approximately similar. We got almost a similar convergence rate of Gibbs sampler for both sets of priors. Here, the analysis for one chain has been exhibited because both the chains have shown equally results with $G(a_1, a_2)$ as before baseline parameters.

The Gelman-Rubin convergence statistic values are equivalent to one. The Geweke test statistic values are somewhat small, and the corresponding p-values are large enough to say that the chains reach stationary distribution. Tables **5-7** contained the posterior mean values and the standard error with 95% credible intervals, the Gelman-Rubin statistics values, and the Geweke test with p-values for Model I, Model II, and Model III. Table **8** included values of AIC, BIC, and DIC values for three models. Table **9** involved values of Bayes factors for Model I, Model II, and Model III.

Values of AIC, BIC, and DIC, given in Table **8**, have been used to compare all models. Model-I holds the lowest possible AIC, BIC, and DIC values, but the values are nearly equal for all models. To decide on model I, Model II, and Model III, we use the Bayes factor. The Bayesian test based on the Bayes factors for Model I against Model II is 6.9508, Model I against Model III is 13.3908, and Model II against Model III is 6.4400, which supports Model I for kidney infection data set compared to other models and frailty is significant in all models. Some patients are expected to be prone to infection compared to others with the same covariate value. It is not surprising, as seen in the data set, there is a male patient with infection time 8 and 16, and there is also a male patient with infection time 152 and 562. Tables **8** and **9** show Model I is better than Model II and Model III.

CONCLUSION

This paper discusses the IG shared frailty model with Weibull exponential, Lomax, and Logistic exponential as baseline distributions. The main aim of our study is to check which distribution fits better. Analysis of kidney infection data has been done in R statistical software with self-written programs. Due to high dimensions, likelihood equations do not show convergence, and it is problematic to apply maximum likelihood estimate. Thus, the Bayesian approach has been used. The Bayesian approach has taken a large amount of computational time, but the time was the same for all three models.

Different prior gives the exact estimates of the parameters. Prior distributions in our proposed models do not affect the convergence rate of Gibbs sampling algorithms. The estimated value of θ (Model-I $\theta = 0.1849$) from the model is very high. It exhibits a strong indication of heterogeneity among the patient in the population for the data set. Bayes factor is used to test the frailty parameter $\theta = 0$, and it is observed that frailty is present. The covariates sex, age, GN, AN, and PKD are the covariates statistically highly impactful for all models. The negative value of the regression coefficient (β_2) of covariate sex indicates that female patients have a slightly lower risk of infection.

With the lowest value of AIC, BIC, and DIC, included in Table **8**, it can be concluded that the first model is more beneficial to use. However, these differences are not much significant. Bayes factor has been used to decide Model I, Model II, and Model III. The model I has been found better. In this case, we can conclude that the IG frailty model with the Weibull exponential baseline distribution is better than Lomax and Logistic exponential baseline distributions. We can also conclude that shared IG frailty with the Weibull exponential distribution as a baseline distribution is a better fit than Lomax and Logistic exponential baseline distributions. By referring to all the above analysis now, we can say that we have suggested a new shared IG frailty model with the Weibull exponential distribution as baseline distribution, which is the best in the proposed models for kidney modeling infection data.

CONSENT FOR PUBLICATION

Not applicable.

CONFLICT OF INTEREST

There is no conflict of interest declared.

ACKNOWLEDGEMENTS

Declared none.

REFERENCES

[1]   C.A. McGilchrist, and C.W. Aisbett, "Regression with frailty in survival analysis", *Biometrics,* vol. 47, no. 2, pp. 461-466, 1991.
        [http://dx.doi.org/10.2307/2532138] [PMID: 1912255]
[2]   D. CLAYTON, "A model for association in bivariate life tables and its application in epidemiological studies of familial tendency in chronic disease incidence", *Biometrika,* vol. 65, no. 1, pp. 141-151, 1978.
        [http://dx.doi.org/10.1093/biomet/65.1.141]
[3]   T. Lancaster, and S. Nickell, "The Analysis of Re-Employment Probabilities for the Unemployed", *Journal of the Royal Statistical Society. Series A (General),* vol. 143, no. 2, p. 141, 1980.
        [http://dx.doi.org/10.2307/2981986]

[4]   J.W. Vaupel, K.G. Manton, and E. Stallard, "The impact of heterogeneity in individual frailty on the dynamics of mortality", *Demography,* vol. 16, no. 3, pp. 439-454, 1979.
        [http://dx.doi.org/10.2307/2061224] [PMID: 510638]

[5]   J. Shih, "A goodness-of-fit test for association in a bivariate survival model", *Biometrika,* vol. 85, no. 1, pp. 189-200, 1998.
        [http://dx.doi.org/10.1093/biomet/85.1.189]

[6]   D. Glidden, "Checking the adequacy of the gamma frailty model for multivariate failure times", *Biometrika,* vol. 86, no. 2, pp. 381-393, 1999.
        [http://dx.doi.org/10.1093/biomet/86.2.381]

[7]   J. Fan, L. Hsu, and R.L. Prentice, "Dependence estimation over a finite bivariate failure time region", *Lifetime Data Anal.,* vol. 6, no. 4, pp. 343-355, 2000.
        [http://dx.doi.org/10.1023/A:1026557315306] [PMID: 11190604]

[8]   P. Hougaard, *Analysis of Multivariate survival Data.* Springer, Verlag: New York, 2000.
        [http://dx.doi.org/10.1007/978-1-4612-1304-8]

[9]   P. Hougaard, "Life table methods for heterogeneous populations: Distributions describing the heterogeneity", *Biometrika,* vol. 71, no. 1, pp. 75-83, 1984.
        [http://dx.doi.org/10.1093/biomet/71.1.75]

[10]   L. Duchateau, and P. Janssen, *The Frailty Model.* Springer: New York, 2008.

[11]   J. Fine, D. Glidden, and K. Lee, "A simple estimator for a shared frailty regression model", *Journal of the Royal Statistical Society: Series B (Statistical Methodology),* vol. 65, no. 1, pp. 317-329, 2003.
        [http://dx.doi.org/10.1111/1467-9868.00388]

[12]   R.S. Chhikara, and J.L. Folks, *The inverse Gaussian distribution.* Marcel Dekker: New York, 1986.

[13]   P. Oguntunde, O. Balogun, H. Okagbue, and S. Bishop, "The Weibull-Exponential Distribution: Its Properties and Applications", *Journal of Applied Sciences,* vol. 15, no. 11, pp. 1305-1311, 2015.
        [http://dx.doi.org/10.3923/jas.2015.1305.1311]

[14]   B. Gompertz, "On the nature of the function expressive of the law of human mortality, and on a new mode of determining the value of life contingencies. In a letter to Francis Baily, Esq. F. R. S. &c", *Philosophical Transactions of the Royal Society of London,* vol. 115, pp. 513-583, 1825.
        [http://dx.doi.org/10.1098/rstl.1825.0026]

[15]   J.G. Ibrahim, M.H. Chen, and D. Sinha, *Bayesian Survival Analysis.* Springer Verlag, 2001.
        [http://dx.doi.org/10.1007/978-1-4757-3447-8]

[16]   S.K. Sahu, D.K. Dey, H. Aslanidou, and D. Sinha, "A Weibull regression model with gamma frailties for multivariate survival data", *Lifetime Data Analysis,* vol. 3, no. 2, pp. 123-137, 1997.
        [http://dx.doi.org/10.1023/A:1009605117713] [PMID: 9384618]

[17]   C.A. Santos, and J.A. Achcar, "A Bayesian analysis for multivariate survival data in the presence of covariates", *Journal of Statistical Theory and Applications,* vol. 9, pp. 233-253, 2010.

[18]   R.E. Kass, and A.E. Raftery, "Bayes Factor", *Journal of the American Statistical Association,* vol. 90, no. 430, pp. 773-795, 1995.
        [http://dx.doi.org/10.1080/01621459.1995.10476572]

CHAPTER 6

An Efficient Approach for Weblog Analysis using Machine Learning Techniques

Brijesh Bakariya*

Department of Computer Science and Engineering, I. K Gujral Punjab Technical University, Punjab, India

Abstract: Information on the internet is rapidly growing day by day. Some of the information may be related to the person or not. The amount of data on the internet is very vast, and it is tough to store and manage. So the organization of massive amounts of data has also produced a problem in data accessing. The rapid expansion of the web has provided an excellent opportunity to analyze web access logs. Data mining techniques were applied for extracting relevant information from a massive collection of data, but now it is a traditional technique. The web data is either unstructured or semi-structured. So there is not any direct method in data mining for it. Here Python programming language and Machine Learning (ML) approach is used from handling such types of data. In this paper, we are analyzing weblog data through python. This approach is useful for time and space point of view because because python has many libraries for data analysis.

Keywords: Data mining, Machine learning, Weblog, Python, World Wide Web

INTRODUCTION

Millions of people use the internet at schools, colleges, offices, houses, and many other places. Moreover, the internet is a standard medium to exchange knowledge from different platforms—web users, *i.e.*, that user who is using the web [1]. Web-users just browse the web and get the information with a single click. But accessing desired information is the most challenging task. There is lots of hidden information present in a single weblog. Moreover, the weblog record contains various kinds of information like IP address, URL, Referrer, Time, *etc.* (Fig. **1**) shows a sample format of the weblog [2]. There are various types of file formats for storing weblog, such as World Wide Web Consortium, Internet Information Services, Apache HTTP Server, *etc.* World Wide Web Consortium is a log of web server that contains a text file with different types of attributes, such as IP address, timestamp, the HTTP

*Corresponding author Brijesh Bakariya:** Department of Computer Science and Engineering, I.K Gujral Punjab Technical University, Punjab, India; Tel: +91-9465884876; E-mail: brijeshmanit@gmail.com.

Krishna Kumar Mohbey, Arvind Pandey & Dharmendra Singh Rajput (Eds.)
All rights reserved-© 2020 Bentham Science Publishers

version, the browser type, the referrer page, *etc*. Moreover, Internet Information Services is also a web server from Microsoft that also contains information about logs with various attributes. There are various types of files including different log file format, but analyzing log and getting information from that log is a very challenging task. The Apache HTTP Server clearly provides log information.

```
199.72.81.55 - - [01/Jul/1995:00:00:01 -0400] "GET /history/apollo/ HTTP/1.0" 200 6245
unicomp6.unicomp.net - - [01/Jul/1995:00:00:06 -0400] "GET /shuttle/countdown/ HTTP/1.0" 200 3985
burger.letters.com - - [01/Jul/1995:00:00:11 -0400] "GET /shuttle/countdown/liftoff.html HTTP/1.0" 304 0
burger.letters.com - - [01/Jul/1995:00:00:12 -0400] "GET /images/NASA-logosmall.gif HTTP/1.0" 304 0
burger.letters.com - - [01/Jul/1995:00:00:12 -0400] "GET /shuttle/countdown/video/livevideo.gif HTTP/1.0" 200 0
205.212.115.106 - - [01/Jul/1995:00:00:12 -0400] "GET /shuttle/countdown/countdown.html HTTP/1.0" 200 3985
d104.aa.net - - [01/Jul/1995:00:00:13 -0400] "GET /shuttle/countdown/ HTTP/1.0" 200 3985
129.94.144.152 - - [01/Jul/1995:00:00:13 -0400] "GET / HTTP/1.0" 200 7074
unicomp6.unicomp.net - - [01/Jul/1995:00:00:14 -0400] "GET /shuttle/countdown/count.gif HTTP/1.0" 200 40310
unicomp6.unicomp.net - - [01/Jul/1995:00:00:14 -0400] "GET /images/NASA-logosmall.gif HTTP/1.0" 200 786
unicomp6.unicomp.net - - [01/Jul/1995:00:00:14 -0400] "GET /images/KSC-logosmall.gif HTTP/1.0" 200 1204
d104.aa.net - - [01/Jul/1995:00:00:15 -0400] "GET /shuttle/countdown/count.gif HTTP/1.0" 200 40310
d104.aa.net - - [01/Jul/1995:00:00:15 -0400] "GET /images/NASA-logosmall.gif HTTP/1.0" 200 786
d104.aa.net - - [01/Jul/1995:00:00:15 -0400] "GET /images/KSC-logosmall.gif HTTP/1.0" 200 1204
129.94.144.152 - - [01/Jul/1995:00:00:17 -0400] "GET /images/ksclogo-medium.gif HTTP/1.0" 304 0
199.120.110.21 - - [01/Jul/1995:00:00:17 -0400] "GET /images/launch-logo.gif HTTP/1.0" 200 1713
ppptky391.asahi-net.or.jp - - [01/Jul/1995:00:00:18 -0400] "GET /facts/about_ksc.html HTTP/1.0" 200 3977
205.189.154.54 - - [01/Jul/1995:00:00:24 -0400] "GET /shuttle/countdown/ HTTP/1.0" 200 3985
ppp-mia-30.shadow.net - - [01/Jul/1995:00:00:27 -0400] "GET / HTTP/1.0" 200 7074
205.189.154.54 - - [01/Jul/1995:00:00:29 -0400] "GET /shuttle/countdown/count.gif HTTP/1.0" 200 40310
ppp-mia-30.shadow.net - - [01/Jul/1995:00:00:35 -0400] "GET /images/ksclogo-medium.gif HTTP/1.0" 200 5866
205.189.154.54 - - [01/Jul/1995:00:00:40 -0400] "GET /images/NASA-logosmall.gif HTTP/1.0" 200 786
ix-orl2-01.ix.netcom.com - - [01/Jul/1995:00:00:41 -0400] "GET /shuttle/countdown/ HTTP/1.0" 200 3985
ppp-mia-30.shadow.net - - [01/Jul/1995:00:00:41 -0400] "GET /images/NASA-logosmall.gif HTTP/1.0" 200 786
ppp-mia-30.shadow.net - - [01/Jul/1995:00:00:41 -0400] "GET /images/MOSAIC-logosmall.gif HTTP/1.0" 200 363
```

Fig. (1). Sample of the weblog.

Machine Learning (ML) is concerned with computer programs that automatically improve their performance through experience. In machine learning, there is a learning algorithm, then data called as training data set is fed to the learning algorithm. The learning algorithm draws inferences from the training data set. It generates a model, which is a function that maps input to the output. Fig. (**2**) shows the process of machine learning. There are various applications of ML, such as Text Categorization, Fraudulent Transactions, Face Recognition, Recommendations, Robot Navigation, Market Segmentation, and many more. There are some important points where learning is used, such as human expertise, does not exist, for example, navigating on mars. Learning is used when humans are not able to explain their expertise; for example, speech recognition [3].

Machine Learning is used when solution changes in time, for example, routing in a computer network. Learning is used when a solution needs to adapt to particular cases; for example, person biometric; there are many areas where learning can be used [4].

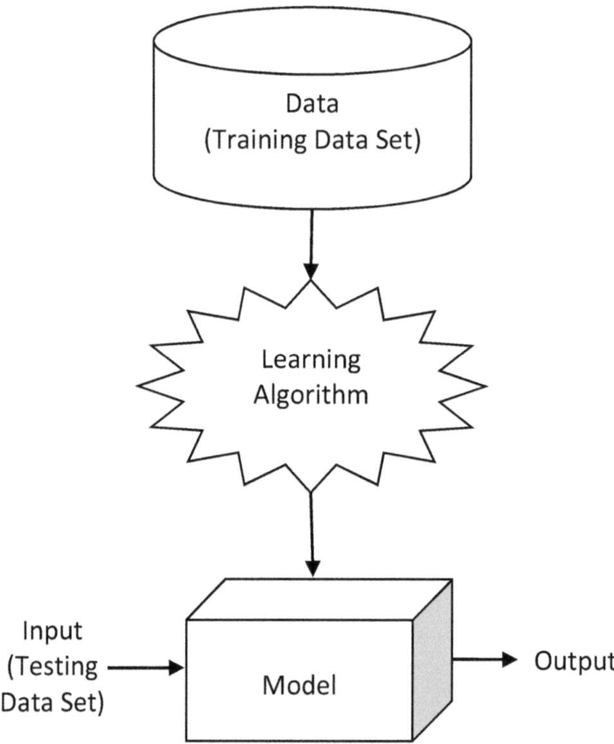

Fig. (2). Process of ML.

MACHINE LEARNING TECHNIQUES

ML algorithms are very efficient for data analysis. In the case of machine learning-based weblog analysis, proposed systems are trained in such using supervised or unsupervised learning algorithms to classify the weblog taken from web servers or other repositories. These are the following types of learning methods:

Supervised Learning

In supervised learning, datasets are divided into two categories, *i.e.*, training datasets and test datasets. In the case of weblog analysis, the training datasets are labeled with IP address, User Name, Timestamp, URL, *etc.* After that, training data sets are used to train the system, and then test data sets are used to test the output of the proposed system.

Unsupervised Learning

In the case of unsupervised learning, labeled training datasets are not used instead of this direct input of datasets that are provided to the system. A relation and features of the inputs are identified in the systems. Once the training phase is completed, then test on test data sets can be performed.

Semi-Supervised Learning

Semi-Supervised Learning is another type of ML approach. It handles both labeled and unlabeled data. This kind of technology system requires a minimum amount of labeled data. So with less effort, more efficient results can be produced using semi-supervised learning.

There are many ML algorithms through this. It can be an analysis weblog, but a few most common and popular machine learning algorithms are used, such as Decision Tree, Support Vector Machine (SVM), Naive Bayes Classifier, Clustering, *etc.* [5, 6]. The decision tree works based on a hierarchical manner. It decides based on features presented in the training data sets. Support Vector Machine is another famous algorithm used in the weblog analysis. It works as a classifier and acts as a line separator. Naive Bayes uses a feature matrix. In this technique, all the features of the feature matrix are considered separately. Naive Bayes classifier is a probabilistic classifier [7]. Clustering is the technique to separate data in different groups of ao clusters. The data point is similar to another one, if it can be separated into one cluster. Similarly, it can make different clusters according to the properties of data.

Python

Python is a high-level, object-oriented programming language. [8]. Nowadays, there is a lot of research work based on python for several reasons because it is a high-level, interpreted, interactive, and object-oriented scripting language [9], [10]. It has many libraries for using data analysis, computation, and processing.

Pandas

Pandas is an open-source python library. This library provides various types of tools data analysis, manipulation, and organization [11, 12]. Moreover, Pandas can make a user-defined function with its inbuilt library [13, 14].

Python with Pandas has various types of applications in different areas such as economics, statistics, finance, science domain, commerce domain, and many more

RELATED WORK

There are various machine learning algorithms and techniques that have been proposed for different kinds of data. We have shown some relevant algorithms for this, A. Tripathy *et al.* [15] used a machine learning algorithm. In this algorithm, they have used Support Vector Machine (SVM) for selecting the best features from the training data. They have compared performance with specific attributes such as recall, f-measure, accuracy, *etc.* The used different datasets and analyzed results according to the dataset. J. Singh *et al.* [16] used four machine learning classifiers, *i.e.*, Naive Bayes, J48, BFTree, and OneR, for optimization of sentiment analysis. They have performed their experiments in three different datasets. They have taken two datasets from Amazon and one dataset from IMDB movie reviews. They compared the efficacies of these four classification techniques (Naive Bayes, J48, BFTree, and OneR) and examined or compared them. They found Naïve Bayes classifier to be quite fast in learning, whereas OneR seems more promising in generating accuracy, precision, and F-measure. G.J. Ansari *et al.* [17] proposed for scene text extraction, recognition, and correction. They used the MSER technique for segmenting text/non-text areas after preprocessing. They have used different models of ML and compare the text data into that model. They also analyzed those data and produced various types of results by using the classification techniques of ML. T. Kucukyilmaz *et al.* [18] developed a model by the use of ML. Such a model shows various types of information. First of all, they preprocess the data and extract the feature from it. The main objective of their model is to enhance performance. After that, they also have experimented with large scale data and measure all the production. Raghavendra T S *et al.* [19] have used classification techniques on tweets. They have to get sentiment from that tweet. Tweets are their data and their sentiments were retrieved. They used the K Nearest Neighbouring (KNN) approach for the extraction of similar tweets. The KNN is an ML algorithm for analyzing the data. They have also designed a framework for pattern extraction concerning tweets logs and producing results with ninety 97% accuracy *via* real-time tweeter logs. The proposed framework is expanded towards pre-processed datasets of major social networking platforms to retrieve higher order of accuracy. The resultant outcomes are processed under reinforcement-based machine learning for web minimization. M.N.M Ibrahim *et al.* [20] developed a strategy for classifying the sentiment of tweets using Naïve Bayes techniques. Positive, negative, or neutral are the three categories of sentiment analysis. They have taken tweets as datasets. They have taken training datasets as tweets and classify the sentiment of tweets of every keyword.

PROPOSED WORK

In this section, we introduce the proposed approach for the analysis of weblog data through python. An algorithm named Weblog Analysis Python (WAP) is also proposed. (Fig. **3**) describes the flow of our proposed approach.

Here we are analyzing weblog data and finding the most visited pages of a website using weblogs. In addition, python and pandas platform were used for accomplishing the task.

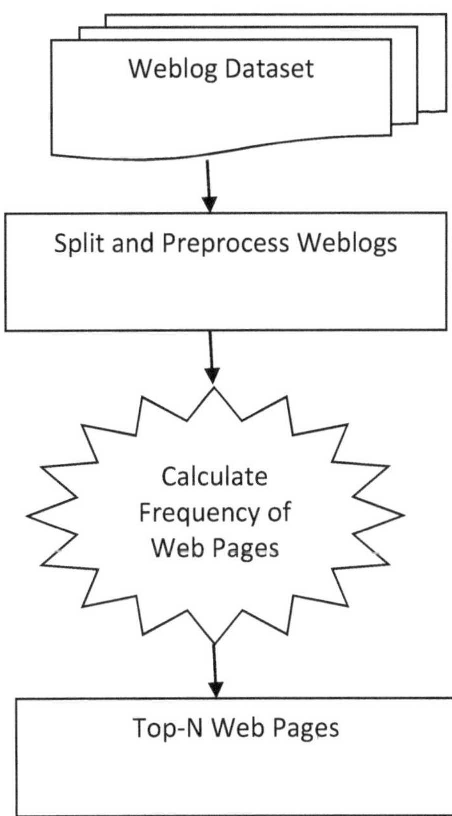

Fig. (3). Flow of the proposed approach.

It has taken the weblog dataset using the apache server log [21]. This log data contains various attributes, such as host identity, user name, timestamp, request status, bytes transferred, referrer, and user agent.

Algorithm:

Weblog Analysis through Python (WAP)

Input:

Weblog Data Wd

Output:

Top-N Pages

1	Read Wd
2	Split Wd and separate every attribute of Wd
3	Preprocess Wd
4	Take one attribute from Wd *i.e.* WebPages (Wp)
5	Count frequency of each Wp
6	Sort all Wp according to its frequency
7	Set the value of N for extracting Top-N page from Wp
8	List the Top-N pages

EXPERIMENTAL RESULTS

The performance of the proposed algorithms WAP is evaluated in this section. The experiment performed on 13257 datasets is downloaded from the Apache Web server log [21]. It has 13257 web records, which consists of IP address, User name, Timestamp, Access request, Result status code, Bytes transferred, Referrer URL, and User-agent. Here we have shown some experimental results. Fig. (**4**) shows a frequency of Top-10 visited web pages.

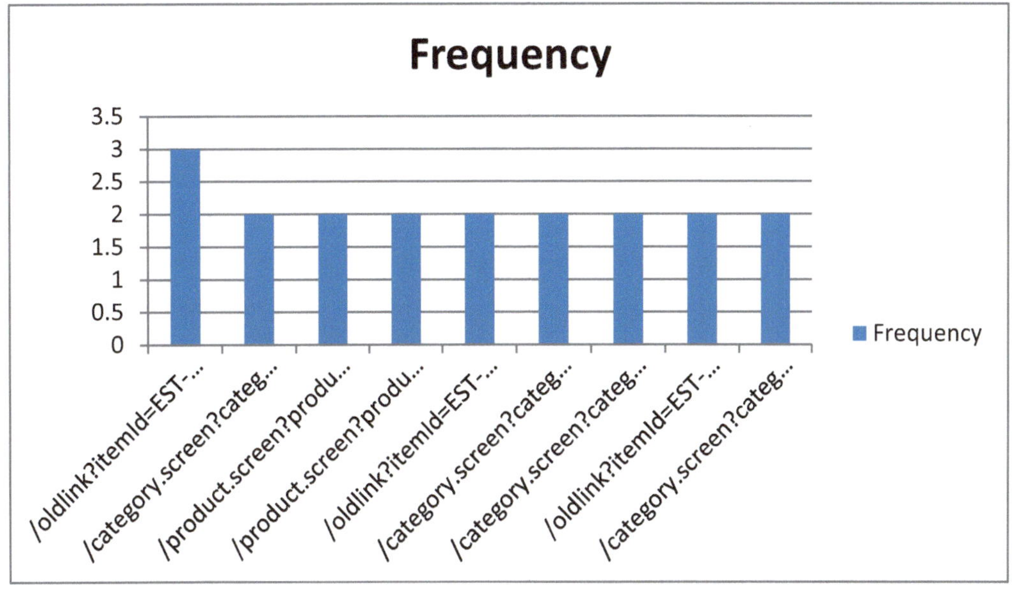

Fig. (4). Frequency of top-10 webpages.

Fig. (**5**) shows the size of weblog datasets based on the frequency. Here the size has decreased after the frequency count. Most of the webpages count as only ones; that is why size has decreased so much. It is very beneficial in terms of memory.

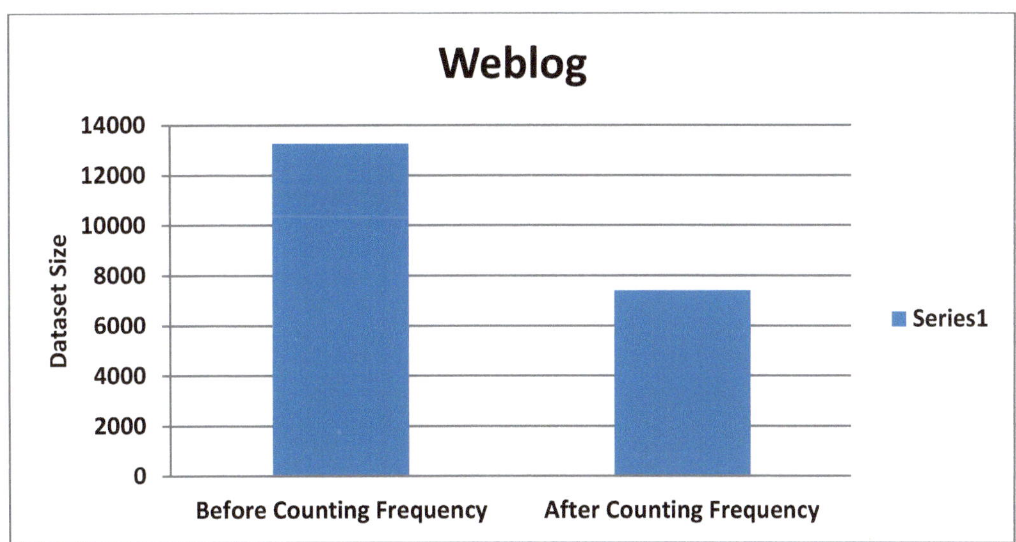

Fig. (5). Dataset size based on the frequency.

CONCLUSION

In this paper, we have proposed an algorithm named Weblog Analysis Python (WAP). Data analysis is a very challenging task within a limited time because the data which is available on the internet is rapidly growing. Millions of websites or e-contents are launching in a single day. That is why the traditional methodology is not compatible with handling such types of massive data. We have used python with pandas library for implementing our algorithm because these types of the library are capable of handling such kinds of extensive data and getting relevant information within limited time and space. Our proposed approach identifies top visited pages and also analyzes the user behavior on the sequences of the requests. The proposed algorithm WAP will be using full advertising and news recommendations.

CONSENT FOR PUBLICATION

Not applicable.

CONFLICT OF INTEREST

There is no conflict of interest declared.

ACKNOWLEDGEMENTS

Declared none.

REFERENCES

[1]   B. Bakariya, and G.S. Thakur, "An Efficient Algorithm for Extracting High Utility Itemsets from Web Log Data", In: *The Institution of Electronics and Telecommunication Engineers (IETE).,* vol. 32. Technical Review, 2015.

[2]   B. Bakariya, and G.S. Thakur, "an Efficient algorithm for extracting infrequent itemsets from weblog", *The international arab journal of information technology,* vol. 16, IAJIT.

[3]   T. Kucukyilmaz, B.B. Cambazoglu, C. Aykanat, and R. Baeza-Yates, "A machine learning approach for result caching in web search engines", *Information Processing and Management,* vol. 8, pp. 834-850.5.
[http://dx.doi.org/10.1016/j.ipm.2017.02.006]

[4]   A.S. Nengro, and K.S. Kuppusamy, "Machine learning based heterogeneous web advertisements detection using a diverse feature set", *Future Generation Computer Systems,* vol. 89, pp. 68-77, 2018.
[http://dx.doi.org/10.1016/j.future.2018.06.028]

[5]   N. Gal-Oz, Y. Gonen, and E. Gudes, "Mining meaningful and rare roles from web application usage patterns", *Computers & Security.,* vol. 82, pp. 296-313, 2019.
[http://dx.doi.org/10.1016/j.cose.2019.01.005]

[6]   N.A.M. Zaib, N.E.N. Bazin, N.H. Mustaffa, and R. Sallehuddin, "Integration of System Dynamics with Big Data Using Python: An Overview", *IEEE Conference,* 2017
[http://dx.doi.org/10.1109/ICT-ISPC.2017.8075337]

[7]   F. Dubosson, S. Bromuri, and M. Schumacher, "A Python Framework for Exhaustive Machine Learning Algorithms and Features Evaluations", *IEEE Conference,* 2016
[http://dx.doi.org/10.1109/AINA.2016.160]

[8]   K. Sahoo, A.K. Samal, J. Pramanik, and S. K. Pani, "Exploratory Data Analysis using Python", *International Journal of Innovative Technology and Exploring Engineering (IJITEE),* vol. 8, October 2019.

[9]   A. Boryssenko, and N. Herscovici, "Machine Learning for Multiobjective Evolutionary Optimization in Python for EM Problems", *In IEEE International Symposium on Antennas and Propagation & USNC/URSI National Radio Science Meeting,* 2018
[http://dx.doi.org/10.1109/APUSNCURSINRSM.2018.8609394]

[10]   N. Rathee, N. Joshi, and J. Kaur, "Sentiment analysis using machine learning techniques on python", *Proceedings of the second international conference on intelligent computing and control systems,* 2018
[http://dx.doi.org/10.1109/ICCONS.2018.8663224]

[11]   R. Filguiera, I. Klampanos, A. Krause, M. David, A. Moreno, and M. Atkinson, "A Python Framework for Data-Intensive Scientific Computing", *IEEE Conference,* 2014
[http://dx.doi.org/10.1109/DISCS.2014.12]

[12]   K.R. Srinath, *Python – The Fastest Growing Programming Language,* vol. 4. International Research Journal of Engineering and Technology, 2017.

[13]   A. Nagpal, and G. Gabrani, "Python for data analytics, scientific and technical applications ", *Amity international conference on artificial intelligence (AICAI),* 2019
[http://dx.doi.org/10.1109/AICAI.2019.8701341]

[14]   A. Kumar, and S.P. Panda, "A survey: how python pitches in it-world", *Amity international conference on artificial intelligence (AICAI),* 2019
[http://dx.doi.org/10.1109/COMITCon.2019.8862251]

[15]   A. Tripathy, A. Anand, and S.K. Rath, "Document-level sentiment classification using hybrid machine learning approach", *Knowledge and Information Systems,* vol. 53, pp. 805-831, 2017.
[http://dx.doi.org/10.1007/s10115-017-1055-z]

[16]   J. Singh, G. Singh, and R. Singh, "Optimization of sentiment analysis using machine learning classifiers", In: *Human-centric Computing and Information Sciences,* vol. 7. 2017.

[17]   G.J. Ansari, J.H. Shah, M. Yasmin, M. Sharif, and S.L. Fernandes, "A novel machine learning approach for scene text extraction", *Future Generation Computing Systems,* vol. 87, pp. 328-340, 2018.
[http://dx.doi.org/10.1016/j.future.2018.04.074]

[18]   T. Kucukyilmaz, B.B. Cambazoglu, C. Aykanat, and R. Baeza-Yates, "A machine learning approach for result caching in web search engines", *Information Processing and Management,* vol. 53, pp. 834-850, 2017.
[http://dx.doi.org/10.1016/j.ipm.2017.02.006]

[19]   T.S. Raghavendra, and K.G. Mohan, "Web Mining and Minimization Framework Design on Sentimental Analysis for Social Tweets Using Machine Learning", *International Conference on Pervasive Computing Advances and Applications, Procedia Computer Science,* 2019
[http://dx.doi.org/10.1016/j.procs.2019.05.047]

[20]   M.N.M. Ibrahim, M. Zaliman, and M. Yusoff, "Twitter Sentiment Classification Using Naïve Bayes Based on Trainer Perception", *In IEEE Conference on e-Learning, e-Management and e-Services (IC3e),* 2015.
[http://dx.doi.org/10.1109/IC3e.2015.7403510]

[21]   https://httpd.apache.org/docs/1.3/logs.html#accesslog.

CHAPTER 7

An Epidemic Analysis of COVID-19 using Exploratory Data Analysis Approach

Chemmalar Selvi G. and **Lakshmi Priya G. G.**[*]

School of Information Technology and Engineering, VIT University, Vellore, India

"Data is the new science. Big Data holds the answers." – By Pat Gelsinger

Abstract: The outbreak of data has empowered the growth of the business by adding business values from the available digital information in recent days. Data is elicited from a diverse source of information systems to bring out certain kinds of meaningful inferences, which serve closer in promoting the business values. The approach used in studying such vital data characteristics and analyzing the data thoroughly is the Exploratory Data Analysis (EDA), which is the most critical and important phase of data analysis. The main objective of the EDA process is to uncover the hidden facts of massive data and discover the meaningful patterns of information which impact the business value. At this vantage point, the EDA can be generalized into two methods, namely graphical and non-graphical EDA's. The graphical EDA is the quick and powerful technique that visualizes the data summary in a graphical or pictorial representation. The graphical visualization of the data displays the correlation and distribution of data before even attempting the statistical techniques over it. On the other hand, the non-graphical EDA presents the statistical evaluation of data while pursuing its' key characteristics and statistical summary. Based on the nature of attributes, the above two methods are further divided as Univariate, Bivariate, and Multivariate EDA processes. The univariate EDA shows the statistical summary of an individual attribute in the raw dataset. Whereas, the bivariate EDA demonstrates the correlation or interdependencies between actual and target attributes; the multivariate EDA is performed to identify the interactions among more than two attributes. Hence, the EDA techniques are used to clean, preprocess, and visualize the data to draw the conclusions required to solve the business problems. Thus, in this chapter, a comprehensive synopsis of different tools and techniques can be applied with a suitable programming framework during the initial phase of the EDA process. As an illustration, to make it easier and understandable, the aforementioned EDA techniques are explained with appropriate theoretical concepts along with a suitable case study.

[*]**Corresponding author Lakshmi Priya G.G.:** VIT School of Design, VIT Vellore, India; Tel: +91-9486322772; E-mail: lakshmipriya.gg@vit.ac.in

Krishna Kumar Mohbey, Arvind Pandey & Dharmendra Singh Rajput (Eds.)

Keywords: Bivariate analysis, Data visualization, Exploratory data analysis (EDA), Multivariate analysis, Statistical methods, Univariate analysis.

INTRODUCTION

Is EDA a Critical Task?

At the outset, the phrase "Data Science" is understood to deal only with statistical data modeling and advanced machine learning techniques. But, the venture of data science stems from the essential keystone, which is frequently underrated or obliterated - Exploratory Data Analysis (EDA). The term EDA was coined by John W. Tukey in 1977 [1]. In the abstract view, EDA is the process of analyzing the main characteristics of the dataset and visualizing the summary of the context of data using statistical methods [2]. It is a critical task before dealing with statistical or machine learning modeling since it perceives the background knowledge required to come about a suitable model to solve the problem at hand and arrive at possible results.

In the early years, computer scientists extracted the knowledge or hidden information from the data by using the technique called Knowledge Discovery Process (KDP). (Fig. **1**)Step 2 of the KDP technique [3] shows data preprocessing where the actual data is cleaned and transformed to make it in a more consistent and understandable format that can be used for inferring knowledge after applying the data mining algorithm.

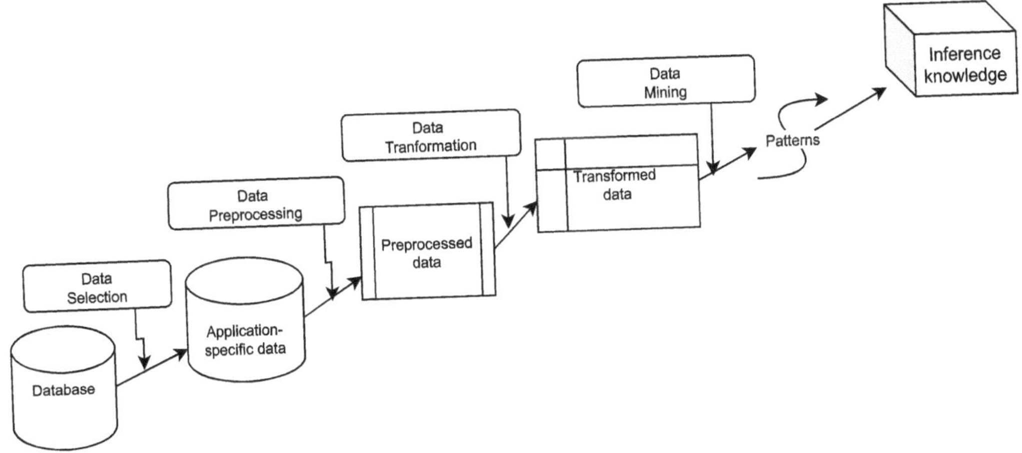

Fig. (1). KDP technique [4].

In recent years, the data scientist unearthed the unknown facts through the EDA process, which is the most important phase of the data science process (Fig. **2**). EDA is highly helpful to the data scientist because of the solutions that they turned out are logically correct and closer to solve the business problems. Apart from turning out with logical solutions, EDA also addresses the business stakeholders by certifying that the right questions are interrogated without ignoring the assumptions and problem statements so as to maximize the benefit of the data scientist's result. This EDA process is a perk to the business stakeholder or data scientist since it provides the more significant intuition behind the data that would not even be thought of to investigate, yet it can be highly demanded insights to the business problems. Hence, EDA is a process or technique which is used to examine the dataset to assess the patterns, identify the relationship, describe the data characteristics, and visualize the statistical summary about the data. It is applied before any data-driven model is constructed.

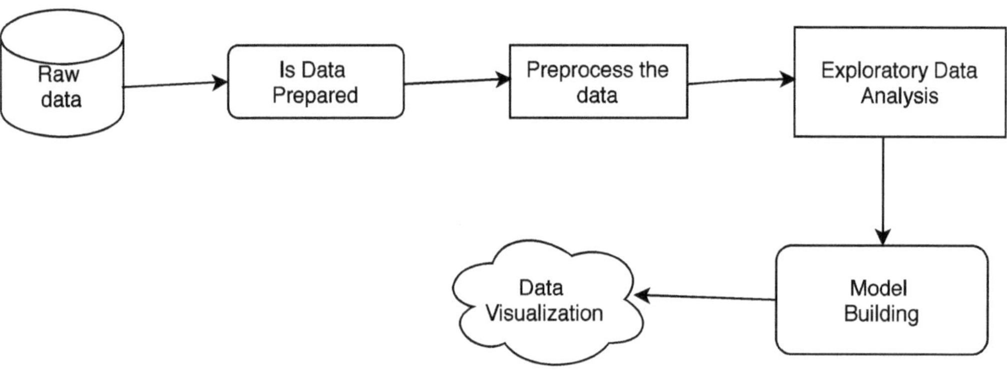

Fig. (2). Data Science Process.

The preface of data analysis is good data exploration and visualization. The EDA ensures the data scientist two-fold:

1. Acquiring more familiarity about the data at hand before developing appropriate business models.
2. Validating the right questions without skewing their assumptions when delivering the results to the business stakeholders.

The following section discusses the methods used in EDA by illustrating the interesting examples wherever suitable.

How Does the Data Scientist Use the EDA?

Data is the fact or information from which the knowledge is inferred. Until the data exist in the world, the EDA process can make one's hand dirty. The data are generally classified into quantitative (or numerical data) and qualitative (or categorical) data [4]. The former type represents the numerical values, which are further grouped as discrete and continuous data (Fig. **3**). The data that can be counted are discrete. For instance, the number of students in a class, the number of workers in a company, and the number of home runs in a baseball game are good examples of discrete data. The data which can be measured is continuous, *e.g.*, the height of the children, the square footage of two bed-room houses, and the speed of the cars. The latter type denotes the categories or labeling the data class and grouped as nominal and ordinal data. The nominal data tags with the class label (For example, Gender, Hair color, ethnicity). On the other side, if the data is ranked or ordered in a certain manner, it is called as ordinal data. Letter grades and economic status are examples of ordinal data. The EDA process is applied to these types of data to refine and define the selection of feature variables that might be appropriate for modeling the machine learning model.

Thus, a better idea to well understand the dataset is to perform multiple data exploratory analysis with which the data scientists gain the confidence to transform the actual dataset into the exciting data for solving business questions. The EDA process is majorly cross-classified as two things. The first is based on the representation of the EDA results, which are graphical or non-graphical. And, the second is based on the number of variables involved in the EDA process, which is either univariate or multivariate [5].

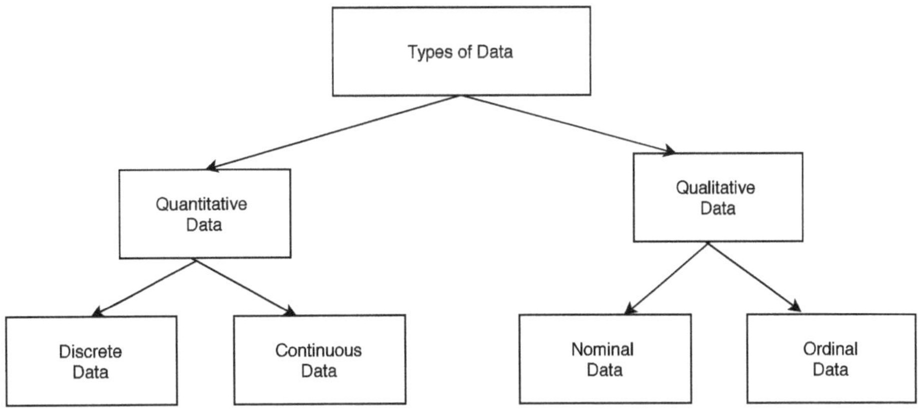

Fig. (3). Types of data.

The non-graphical EDA generally performs the numeric calculation to show the statistical summary of the large dataset while the graphical EDA presents the complete summary (A picture with a thousand words) of the dataset diagrammatically. On the other side, univariate EDA involves one variable only (one column data), and multivariate EDA involves two or more variables normally to look at the relationship between these variables. Here, the presented multivariate EDA will be bivariate EDA (exploring mostly two variables), but at times it may also involve more than two or three variables and so.

To have a better understanding of EDA methods, further subsections explain the EDA concepts alongside the new case study carried out with the COVID-19 dataset, which is publicly available for researchers and academicians collated by John Hopkins University [6, 7]. Using python jupyter notebook, the EDA methods are discussed. The sample of the COVID-19 dataset is shown in Fig. (**4**), which has 25676 rows and 8 columns.

Univariate EDA Methods

The simplest and easiest way of data analysis is the univariate analysis, which uses only one variable in it. Because it is using only one variable, it is incapable of showing the relation or purposes [8, 11]. The motive of univariate data analysis is to uncover the hidden patterns by simply describing the statistical summary of the data. First, the non-graphical EDA methods are discussed.

	State	**Country**	**Lat**	**Long**	**Date**	**Confirmed**	**Deaths**	**Recovered**
0	NaN	Afghanistan	33.000	65.000	1/22/2020	0	0	0
1	NaN	Albania	41.153	20.168	1/22/2020	0	0	0
2	NaN	Algeria	28.033	1.659	1/22/2020	0	0	0
3	NaN	Andorra	42.506	1.521	1/22/2020	0	0	0
4	NaN	Angola	-11.202	17.873	1/22/2020	0	0	0

Fig. (4). COVID-19 sample dataset.

Descriptive Statistics

Descriptive Statistics is the description that summarizes the dataset with either the entire or sample of the dataset. The descriptive statistics can be performed by measures of central tendency (mean, median, and mode), measures of variability (spread), and skewness and kurtosis. The measures of central tendency calculate the mean, median, and mode. The measures of variability find the standard deviation, variance, minimum and maximum values, while the skewness and kurtosis describe the distribution of data. Thus, the descriptive statistics are applied to quantitative data, which summarize it, as shown in Fig. (**5**).

	Lat	Long	Confirmed	Deaths	Recovered
count	25676.00000	25676.00000	2.567600e+04	25676.00000	25676.00000
mean	21.433571	22.597991	2.621362e+03	164.070455	664.519707
std	24.740902	70.570871	2.524682e+04	1710.575849	5592.498342
min	-51.796300	-135.000000	0.000000e+00	0.000000	0.000000
25%	7.000000	-19.020800	0.000000e+00	0.000000	0.000000
50%	23.659750	20.921188	7.000000e+00	0.000000	0.000000
75%	41.204400	81.000000	2.030000e+02	2.000000	21.000000
max	71.706900	178.065000	1.012582e+06	58355.000000	123903.000000

Fig. (5). Quantitative descriptive statistics.

Box Plot

Box plot is the graphical representation of quantitative data using box and whiskers. It is an effective way of summarizing the data and identifying the outliers as well. The box plot consists of five-number summary: minimum, first quartile (25%), second quartile (median or 50%), third quartile (75%), and maximum.

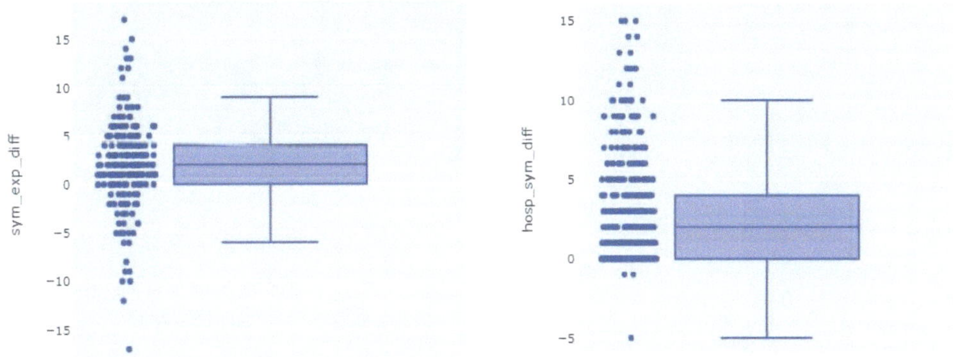

Fig. (6). (a) Days difference between a symptom and exposure dates. **(b)** Days difference between a hospital visit and symptom dates.

The above figures show the box plot where Fig. (**6a**) infers that for the majority of the people, the symptoms of coronavirus seems to be exposed only after 0 or more days. And Fig. (**6b**) tells that the exposed people with the possible symptom of coronavirus visit to the hospitals only after a few days, which is the major cause for the rapid spread of the infected disease.

Histogram

The histogram is one of the best ways to visualize qualitative data. The bar chart or histogram is the simple graph with some frequency or proportion (count) with a certain range of values. The histogram (Fig. **7)** shows the count of COVID-19 cases recorded as three categories, namely: Recovered, Deaths, and Confirmed. Also, it presents the top 10 countries in the world, which is more affected by which the United States (US) stands as the first and most affected country in worldwide.

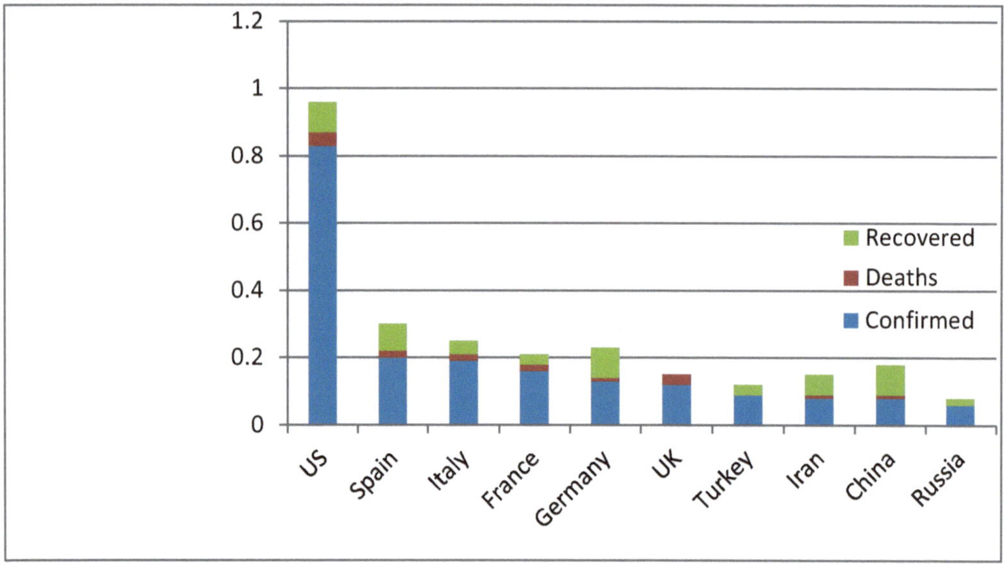

Fig. (7). Top 10 Country-wise COVID-19 cases and its consequences.

MULTIVARIATE EDA METHODS

Multivariate EDA methods [9, 11] are used to examine multiple variables at the same time to exhibit the relationship between them.

Cross-Tabulation

Cross-tabulation is one of the primary bivariate data analysis, which is non-graphical. It can be applied to both quantitative and qualitative data with a minimum number of variables. For instance, if two variables are used, then a two-way table is built with column and row heading corresponding to the variable names. In cross-tabulation (Fig. **8**), is presented to show the different categories of COVID-19 cases reported against each country.

	Confirmed	Deaths	Recovered	Active
Country				
USA	1133069	66385	175382	891302
Spain	216582	25100	117248	74234
Italy	209328	28710	79914	100704
United Kingdo m	183500	28205	901	154394
France	168518	24763	50663	93092
Germany	164967	6812	130600	27555
Russia	134687	1280	16639	116768
Turkey	124375	3336	58259	62780
Iran	97424	6203	78422	12799
Brazil	97100	6761	40937	49402
China	83959	4637	78586	736
Canada	57927	3684	23814	30429
Belgium	49906	7844	12309	29753

Fig. (8). Country-wise COVID-19 reported cases.

Correlation Matrix

The correlation matrix is another multivariate non-graphical EDA method that is used to display the level of relationship that exists between two or more interdependent variables. The matrix projects the numeric value ranging from 0 to 1. If the matrix cell represents 0, it is a weak correlation, and 1 means a strong correlation. For instance (Fig. **9**), the cases recorded as 'Confirmed' exhibit a strong correlation with 'Active' and weak correlation with 'Recovered.'

	Confirmed	Deaths	Recovered	Active
Confirmed	1.000000	0.928797	0.804834	0.980741
Deaths	0.928797	1.000000	0.780704	0.887756
Recovered	0.804834	0.780704	1.000000	0.675170
Active	0.980741	0.887756	0.675170	1.000000

Fig. (9). Correlation matrix for COVID-19 country-wise data.

Maps

Maps are the graphical EDA method, which is a different kind of visual data projecting more amount of information. The most popular map used in the data visualization method is the choropleth map, which uses a different color and various sizes to highlight the importance of essential information. In the given map (Fig. **10**), the red color circles highlights the COVID-19 affected areas, and the size of the circle shows how much the region is affected by the disease.

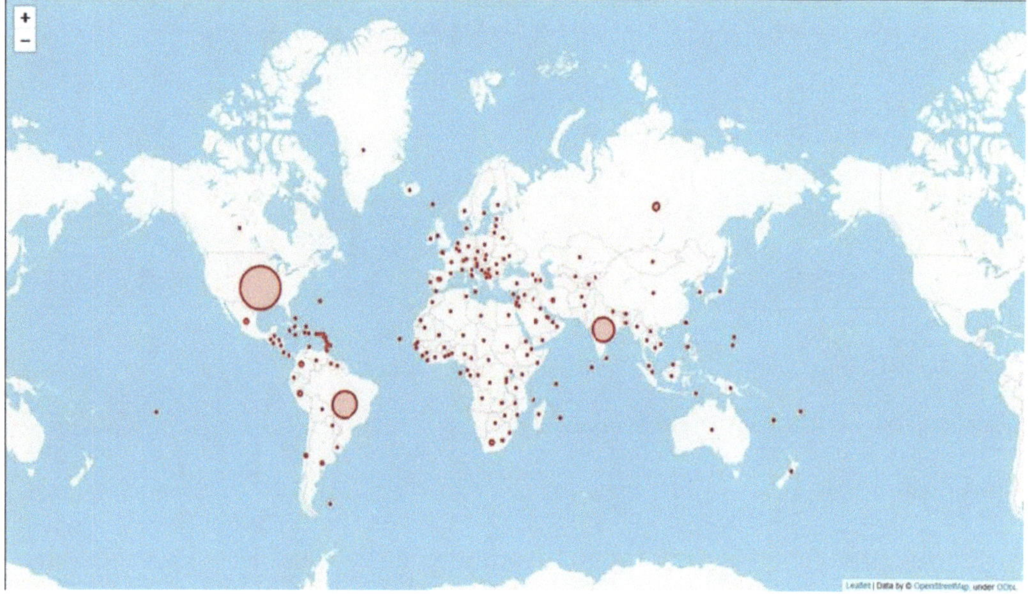

Fig. (10). Map showing COVID-19 affected region.

Graphs

Graphs are the eye-catching multivariate graphical EDA method. It is the most vibrant data visualization method that shows the data in a multi-dimensional axis with a sequence of coordinates. Each coordinate is the value displaying information, which is a particular mathematical or time-stamp based relationship. For example (Fig. **11**) presents the percentage of COVID-19 patients recorded as a positive case for each day. The graph visually shows the rapid growth of the total number of positive cases increasing concerning the time.

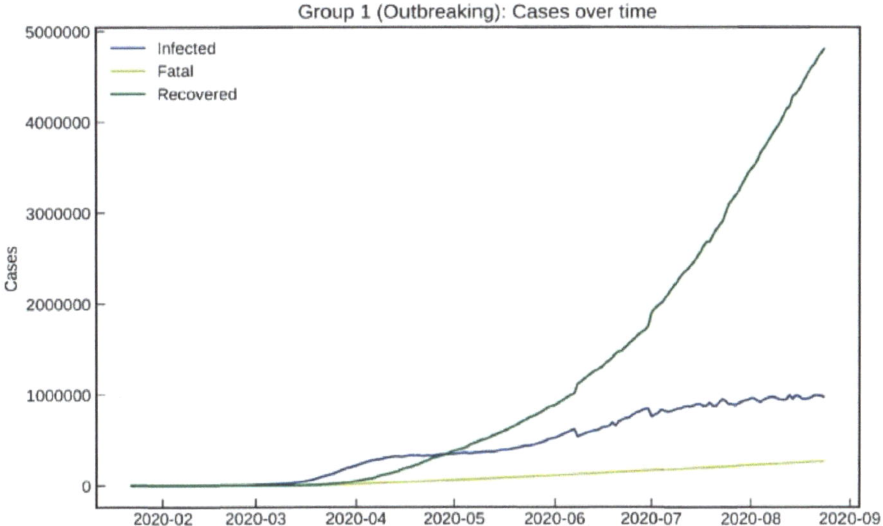

Fig. (11). Positive Cases (%) *vs.* Time.

In the next section, the tools or platforms supporting the EDA mentioned above methods are discussed, which helps the reader to understand the ways of exploring and visualizing the data.

DOES PROGRAMMING KNOWLEDGE REQUIRED IN THE EDA PROCESS?

To perform data exploration and visualization, knowledge of coding is not a top priority. There exist a plethora of open source tools or platforms [10] which provides automated steps in data analysis like dean cleaning, data preprocessing, data visualization, and so on. Table **1** lists the non-programming EDA tools which are freely available for data exploration and analysis, and very few of which are licensed actually.

Table 1. Non-programming EDA tools.

Tool	Description
Excel/Spreadsheet	Popular and powerful tool for data wrangling, data analysis, and visualization.
Trifacta	Intelligent tool with incredible features supporting data wrangling operations.
RapidMiner	Fascinating tool for advanced data analysis and model building.
Rattle GUI	It is closer to R, which includes many robust machine learning algorithms like SVM, Neural networks, and so on.
Qlikview	It is an excellent business intelligence tool that provides enough data exploration, trends, and insights about the complete data.
Weka	A machine learning tool that is faster and easier to build models and deploy it.
KNIME	It is used as a data analytics tool which supports various advanced machine learning algorithms.
Orange	Interactive data mining and data visualization tool.
OpenRefine	Early as Google Refine, performs predictive modeling.
Talend	A data collaboration tool used for decision making in business.

With this handful list of EDA tools, there are many other programming languages which are effectively used by the computer programmers and data analyst to gain better insight about the data at hand. The most popular and simple programming languages, R and Python, are increasingly used as general approaches for data exploration and analysis. Hence, EDA is the most crucial and the most important phase in the data science process.

PROTOCOL GUIDING WHEN AND WHERE EDA IS EFFICIENT

EDA is an important part of the analytical phase in any data science project. It helps to make sense of the data even before the first model is built [9]. By this phase, the better understanding of the data, getting insight into the data patterns, identifying hidden relations between the data, and detecting the outliers or examining the anomalies can be well learned.

The highlights of EDA methods are:

1. Identifying patterns and validating the assumptions.
2. Learning the data instinct.
3. Consolidating data exploration with visual effects.

CONCLUSION

EDA seems to be a similar terminology as statistical modeling, but they are not identical. EDA is a general approach used for data analysis, which is mostly a graphical approach that includes different ways of uncovering underlying assumptions, validating the hypothesis, and extracting the hidden features by exploring the data from all sides. It is a state-of-the-art technology used for data visualization and for familiarizing the data at hand. In this chapter, the methods used in the EDA technique are discussed along with the interesting COVID-19 case study. The chapter also explains the ways and significance of EDA techniques in the data science process. Thus, before any model is built for the business problem, the EDA is to be performed to highly ensure that the data at hand is appropriate to develop the business model.

CONSENT FOR PUBLICATION

Not applicable.

CONFLICT OF INTEREST

There is no conflict of interest declared.

ACKNOWLEDGEMENT

Declared none.

REFERENCE

[1]   J.W. Tukey, *Exploratory data analysis.* Addison Wesley, 1977.

[2]   J. de Mast, and B.P. Kemper, "Principles of exploratory data analysis in problem solving: what can we learn from a well-known case?", *Quality Engineering.,* vol. 21, no. 4, pp. 366-375, 2009.
[http://dx.doi.org/10.1080/08982110903188276]

[3]   G.J. Myatt, *Making sense of data: a practical guide to exploratory data analysis and data mining.* John Wiley & Sons, 2007.
[http://dx.doi.org/10.1002/0470101024]

[4]   J. Han, J. Pei, and M. Kamber, *Data mining: concepts and techniques.* Elsevier, 2011.

[5]   C.H. Yu, "Exploratory data analysis in the context of data mining and resampling", *International Journal of Psychological Research,* vol. 3, no. 1, pp. 9-22, 2010.
[http://dx.doi.org/10.21500/20112084.819]

[6]   https://github.com/datasets/covid-19.

[7]   S.K. Dey, M.M. Rahman, U.R. Siddiqi, and A. Howlader, "Analyzing the epidemiological outbreak of COVID-19: A visual exploratory data analysis approach", *Journal of Medical Virology,* vol. 92, no. 6, pp. 632-638, 2020.
[http://dx.doi.org/10.1002/jmv.25743] [PMID: 32124990]

[8]   Y. Tsumoto, and S. Tsumoto, "Exploratory univariate analysis on the characterization of a university hospital: A preliminary step to data-mining-based hospital management using an exploratory univariate analysis of a university hospital", *The Review of Socionetwork Strategies,* vol. 4, no. 2, pp. 47-63, 2010.
[http://dx.doi.org/10.1007/s12626-010-0014-x]

[9]   K. Andreas, "A survey of methods for multivariate data projection, visualisation and interactive analysis", *Proceedings of the 5th International Conference on Soft Computing and Information/ Intelligent Systems,* 1998 pp. 55-59

[10]   A. Ghosh, M. Nashaat, J. Miller, S. Quader, and C. Marston, "A comprehensive review of tools for exploratory analysis of tabular industrial datasets", *Visual Informatics,* vol. 2, no. 4, pp. 235-253, 2018.
[http://dx.doi.org/10.1016/j.visinf.2018.12.004]

[11]   J. Heer, M. Bostock, and V. Ogievetsky, "A tour through the visualization zoo", *Communications of the ACM,* vol. 53, no. 6, pp. 59-67, 2010.
[http://dx.doi.org/10.1145/1743546.1743567]

SUBJECT INDEX